Explorations in Ecocriticism

Ecocritical Theory and Practice
Series Editor: Douglas A. Vakoch, California Institute of Integral Studies, USA

Ecocritical Theory and Practice highlights innovative scholarship at the interface of literary/cultural studies and the environment, seeking to foster an ongoing dialogue between academics and environmental activists. Works that explore environmental issues through literatures, oral traditions, and cultural/media practices around the world are welcome. The series features books by established ecocritics that examine the intersection of theory and practice, including both monographs and edited volumes. Proposals are invited in the range of topics relevant to ecocriticism, including but not limited to works informed by cross-cultural and transnational approaches; postcolonialism; posthumanism; ecofeminism; ecospirituality, ecotheology, and religious studies; film/media and visual cultural studies; environmental aesthetics and arts; ecopoetics; ecophenomenology; ecopsychology; animal studies; and pedagogy.

Recent Titles in the Series

Feminist Ecocriticism: Environment, Women, and Literature, eited by Douglas A. Vakoch

Ecoambiguity, Community, and Development: Toward a Politicized Ecocriticism, edited by Scott Slovic, R. Swarnalatha, and Vidya Sarveswaran

Transversal Ecocritical Praxis: Theoretical Arguments, Literary Analysis, and Cultural Critique, by Patrick D. Murphy

Urban Ecologies: City Space, Material Agency, and Environmental Politics in Contemporary Culture, by Christopher Schliephake

Ecocriticism of the Global South, edited by Scott Slovic, R. Swarnalatha, and Vidya Sarveswaran

Explorations in Ecocriticism: Advocacy, Bioregionalism, and Visual Design, by Paul Lindholdt

Explorations in Ecocriticism

Advocacy, Bioregionalism, and Visual Design

Paul Lindholdt

LEXINGTON BOOKS
Lanham • Boulder • New York • London

Published by Lexington Books
An imprint of The Rowman & Littlefield Publishing Group, Inc.
4501 Forbes Boulevard, Suite 200, Lanham, Maryland 20706
www.rowman.com

Unit A, Whitacre Mews, 26-34 Stannary Street, London SE11 4AB

British Library Cataloguing in Publication Information Available

Library of Congress Cataloging-in-Publication Data

Lindholdt, Paul J.
Explorations in ecocriticism : advocacy, bioregionalism, and visual design / Paul Lindholdt.
pages cm. — (Ecocritical theory and practice)
Includes bibliographical references and index.
ISBN 978-0-7391-9498-0 (cloth) -- ISBN 978-0-7391-9499-7 (electronic)
1. Ecocriticism. I. Title.
PN98.E36L56 2015
809'.9336--dc23
2015000472

™ The paper used in this publication meets the minimum requirements of American National Standard for Information Sciences Permanence of Paper for Printed Library Materials, ANSI/NISO Z39.48-1992.

Printed in the United States of America

Contents

Introduction

This book explores the range of ecocriticism. Unlike related books of literary or culture criticism, this one hybridizes my scholarship with other forms of writing, chiefly journalism and creative nonfiction. Our genres are more porous than we acknowledge. Creative writers may rely on weighty research, journalists aim to make their reporting as creative as editors allow, and scholars find pride in persuading one another through the sound and the sense of their completed research. The idea for greater hybridity gained legitimacy for me when an appraisal in *Hudson Review* characterized my 2011 ecological memoir as ecocriticism—a project with a critical value and design.

Prepared as an early Americanist, I designed my terminal degree to green my research. I hoped to regain a sense of consequentiality that I found lacking in academe—a topic I explore at length here in "Literary Activism and the Bioregional Agenda" and in "Restoring Bioregions Through Applied Composition." Beyond impacts in the classroom, I wondered in frustration, what could we hope for by way of academic outcomes? In my doubt and dismay, I found inspiration within environmental education and ecocriticism. The generous and supportive folks who labor in those fields renewed my interest and my faith in pushing through a research agenda, an agenda that had been in serious jeopardy.

Under Harrison Meserole, former MLA Chief Bibliographer, an eminent mentor, I learned the ropes of American literature and wrote a dissertation on America's earliest natural historian, John Josselyn. Some of that research spills over into discussions here titled "Edging Toward Ecology in Early American Natural Literature" and "West of Winthrop: Language and Landscape in Washington Territory." Once the University Press of New England issued the Josselyn project as a book, I became acquainted with innovative and inspiring scholars who pioneered new journals and invited me to join their efforts—first as member of the editorial advisory board of *ISLE: Interdisciplinary Studies in Literature and Environment*, then as a charter board member of the online *Journal of Ecocriticism*. Michael Branch, one of the founders of the Association for the Study of Literature and Environment, proved particularly nurturing as an editor and friend whose preparation and interests in textual editing, early Americana, and journalism parallel my own.

As a junior professor I found my sustenance in the ecological, but I also hungered to expand the boundaries of the emerging ecological criticism, to nudge it beyond the purely literary and into a culture criticism. Participating in American studies conferences helped me manage that transition. Purposefully I swerved my work into anthropology, environmental history, politics, rhetoric, visual arts, and other neglected areas where I could elaborate more freely. Such swerves ribbon through the present book. Deliberately I aimed to expand the range of the disciplines, to tug them from the furrows of arcana and specialization. Also I aimed to apply original approaches, including imperialist nostalgia and ecopornography (afflicted though the latter concept is with excessive semantic noise), in the essays "West of Winthrop" and "Overtures to Sublimity." The fine arts play integral parts in both of those discussions, as they do in "An Iconography of Sabotage."

Besides striving for more breathing room within the interdisciplinary, I worked to personalize some discourse forms, as in the hybridized pedagogy and memoir that ground "Gifts and Misgivings in Place." Gone forever are leisurely treatments of literary eras by the likes of Vernon L. Parrington and Van Wyck Brooks, whose books relied on atmospheric renderings of time and place. The personal voice in scholarship has been often lacking in the generations since the writings of those guiding lights. But readers might begin to wish again for some pleasing medium between the two extremes—between criticism written by imaginatively recreating scenes, and criticism undertaken as if it were a kind of science. Ecocriticism commendably has escaped both blights so far.

In such practical concerns I am fortunate to have had Annie Dillard as a mentor. After she won the nonfiction Pulitzer Prize for *Pilgrim at Tinker Creek*, she relocated to an island on Puget Sound and commuted to the mainland, where I took her classes at Fairhaven College. As a teacher she was a wild ride. She chain-smoked while she taught, directed us to stamp out metrical time to sonnets by Gerard Manley Hopkins, and rightly convinced me to toss my earliest creative writing to break free of its spell. As a writer she was an equally powerful influence for her students. Flagrantly independent as a naturalist and a thinker, brilliantly incisive in person, she grieved what we lose when we vanquish our connections to nature, when we overlook its minutiae, and when a complacent guise of objectivity masquerades as Truth. Her living example propelled me into more intimate discourse forms, a practice that I believe to be entirely consistent with sound scholarship.

An interest in radical activism also charges this book. Statutes of limitation have by now run their course, and so I credit Earth First! for shaping my early consciousness. In Bellingham in the 1980s a ragtag bunch of us gathered weekly to rant and strategize, organize and plan. We practiced a form of potlatch, in which we gave away our fears for the future of the world we lived in. Those were the heady days before Dave Foreman was busted and Edward Abbey was still alive. In Pocatello, Idaho, in 1984, Abbey harangued a standing-room-only crowd about the ecological devastation of cattle grazing. And while I recall no gunfire outside the lecture hall that night, as he said took place in other venues, the atmosphere was tense and jubilant for all of us who had been witnesses to the waste of the West at the behest of ranchers and their agency retainers. Abiding interests in environmental politics rise to the surface throughout this book, but particularly in "An Iconography of Sabotage" and "Rage Against the Machine."

Countering my interest in radicalism in these pages is a hopeful reverence. Within the best of ecocriticism, I always appreciate devotion for the Earth, although one might feel apologetic for capitalizing the name of this home globe as we do for other planets. Devotion or reverence, manifested as "sense of place" in some writings, also "topophilia" or a bioregional ethos, is consistent with all the major world religions, or so both Wendell Berry and Thomas Berry have argued and demonstrated in their walks as Christians. Such manifestations of respect or veneration I excavate, along with nascent ecological insights during the colonial era, in "Edging Toward Ecology in Early American Natural History."

For reasons difficult to fathom, many on the Christian Right choose to vilify rather than valorize those who would praise the creation and struggle to protect it. As early as 1970—the year of the first Earth Day, when John McPhee was completing field research for *Encounters with the Archdruid,* his book about Friends of the Earth founder David Brower and Brower's antagonists—skeptical regard for environmentalists was fueling the emerging culture wars. In McPhee's 1971 book, developer Charles Fraser decries environmentalists for the ways they "worship trees and sacrifice human beings to those trees," thus resembling ancient pagan druids. Backlash to environmentalism is my topic in "American Nature Writing and the Wise-Use Movement." These studies are meant as hallowed manifestoes or sacred screeds for the ways they promote and praise a biophilia—an innate affinity in humans for other life forms, a passion for all life and the living world, a philosophy that remarkably still threatens many people of faith.

Like Thomas Berry and Wendell Berry, I credit revelations. So, evidently, does Thomas Pynchon. In an essay he wrote on the Luddites in 1984, the former engineering major declared, "To insist on the miraculous is to deny to the machine at least some of its claims on us, to assert the limited wish that living things, earthly and otherwise, may on occasion become Bad and Big enough to take part in transcendent doings." What a clever phrase, "transcendent doings," especially in its confidence that living things like dead things might partake of the miraculous, that living things might capably countervail the dominance of machinery in our lives. Pynchon offered a sorely needed sanctuary for the sacred, a proxy for technology and dead martyrs, even if his was only wishful thinking.

Pynchon's skeptical take on technology also informs several discussions here. My research into Edward Abbey and his fans builds upon an appreciation of Luddism and its latter-day descendant neo-Luddism to appreciate anew those dire forces of modernity that sent Abbey into such fiery flights of ridicule and rage. Had Abbey and Pynchon met, they might have gotten along. Technology's downside is inescapable in my discussions here of dams in "Overtures to Sublimity" and "An Iconography of Sabotage." Technocracy, to be precise, not technology, is what impinges on our beings and complicates our vision of this home planet—technocracy as a belief in the fit ascendency of technical experts.

Without a sound foundation in nature and natural history, I would not have kept my interests in the ecological alive to feed my calling inside higher education. For that foundation I credit my father. Without sanctifying him, I can say that he led me to the blessed water while never compelling me to drink. He took me into field and stream and never obliged me to squeeze the trigger or set the hook. And that perhaps is all we can do, as I assert in "Gifts and Misgivings in Place"—bundle our selves and our progeny into the great outdoors to kindle the wonder that life in modern times might stifle. Here in the discussion "West of Winthrop" I release some historical indignation over the seizure of the nature-writing mantle by sick souls, in William James's insightful phrase, those faux transcendentalists who exalt Emerson's not-me while depreciating their fellow beings.

I wrote original versions of these essays from 1995 through 2009. When I began to review them for this book, I found myself reworking them much more than I had at first intended; work expands to fill available time for its completion. While I was updating my research and revising my earlier opinions, I kept an eye on the need to integrate the discussions as fully as possible. Also I found some interpretations evolving over the years, for instance in my regard for our earliest natural historians, at whom most literary critics tend to sniff. Moreover, I wanted this book to advance my current views, not those of the past. As the spirit of revision gripped me, I began adopting approaches from other disciplines and composing entirely new passages and one new chapter.

Among the many people I wish to thank here, I include Kathleen Dean Moore, whose activism as a philosopher has been such a positive model for all of us who aspire to do public-intellectual groundwork. Robert Michael Pyle, with his eclectic interests and generosity to fellow writers, has been inspiring. My wife Karen Palrang Lindholdt also managed to bestow on me the gift of time while sustaining her legal career and co-parenting our two boys. Scott Slovic and Rebecca Raglon, for their selfless drudgery in respectively editing *ISLE* and *Journal of Ecocriticism*, will find their names etched upon the celestial spheres. Also I wish to thank Margaret Zoller and Marisa Bourgoin, Chief of Reference Services, Smithsonian Archives of American Art; John R. Waggener, Associate Archivist, American Heritage Center, University of Wyoming; and Gloria Mestas and Brit Story, late of the Bureau of Reclamation. Tony Flinn and

Paul Metzner kindly read portions of this book in manuscript, and Pierre Lagayette furnished the spur for some of this work with the stimulating symposia he sponsored early in this millennium at the University of Paris-Sorbonne. Finally, credit must go to Eastern Washington University and its Northwest Institute for Advanced Study, which awarded me several professional leaves and grants during my two decades in residence.

ONE

Edging Toward Ecology in Early American Natural History

Few enough readers have shown interest in early American natural histories—accounts of nature in North America between 1492 and 1800. The genre is tainted by religious and commercial impulses, afflicted by racial suppositions, and therefore even ecocritics have neglected to turn much attention to these disparate and non-canonical texts. If read as science-in-the-making, however, such quaint books can enchant with their fresh views of a strange new world. They can inspire with the primacy of their power to chronicle great changes in the land. As incipient science, they demonstrate an ongoing conflict between the dictates of rational order, on the one hand, and the abiding mysteries of nature on the other. As statements of interpretive ecology, early natural histories foreshadow many of the same sustainability issues that perplex us *Homo sapiens* today.

As I use it in this discussion, "interpretive ecology" denotes not only a process of remarking ecological relations, but also assessing causes and effects of those relations. The word *ecology,* says the *Oxford English Dictionary* (*OED*), originated not until 1876 when one E. R. Lankester listed it as among the "great series of phenomena" and spelled it *œcology*. As a principle for apprehending the natural world, of course, ecology predated that early usage, when humans studied "the relationships between living organisms and their environment." Aside from the delight that might derive from seeing science in its infancy, instances of early ecology are challenging to identify. High emotions and sublime ecstasies compound the ecological observations and make them inconsistent with scientific standards of excellence today. Early natural historians exploited nature, brash in their certitude that the rightful place of white Europeans was atop the food chain. Boosterism and its hyperbole further afflicted the sought-after objectivity that was already believed to typify sound science. Early natural histories, then, inhabit a fertile but uncertain nether-realm that resides fitfully between scientific fact and artistic license. Here I strive to make some taxonomic sense of these works, vis-à-vis the secondary literature, and I survey some of the prominent primary texts to provide examples. In the interest of keeping the varied taxonomic distinctions from distracting, and to diminish undue semantic noise, I recur to the term "natural history" as an interim catchall phrase, despite its relative imprecision.

These early documents have suffered from three strikes. First, as Greg Garrard noted, "nature writing, which enjoys considerable popularity in the USA, has tended to be downgraded by academic prejudices favouring fiction over non-fiction, and human dramas over narratives of interaction of humanity and nature" (58). Second, interest in colonial literature and culture has yielded in recent years to theoretical post-colonialism, an educational trend that tends to diminish natural histories to irrelevance. The index for Garrard's book, as an instance, has no entries for "colonialism"; it directs readers instead to entries for "imperialism" and "ecocriticism: postcolonial," effectively synonymizing the imperial and the colonial, the postcolonial and the colonial. As a third strike against early American natural history, academic disputes confirm that even those few canonical texts, William Bartram's 1791 *Travels* for example, are fraught with controversy between the few scholars who still care to treat them. In spite of the obstacles such texts pose for ecocritics, I side with Garrard and Harrison, who hold "that colonial scientists, working in the 'natural laboratories' of tropical and subtropical islands, laid the foundations of modern environmentalism by discovering that nature could be *fragile*" (Garrard 194). Indeed, the avian species discussed below, the dodo, the auk, the cahow, the passenger pigeon, went from plentiful to extinct, or undiscoverable for centuries like the cahow.

Ecocritic Christoph Irmscher aimed to soften such strikes against the genre. In his *Poetics of Natural History* (1999), he noted for instance that William Hedges ranked Bartram's *Travels* "'the most astounding verbal artifact of the New Republic'" (36), even though Irmscher himself found the Bartram book "strange": a "ragbag of different and apparently unreconcilable genres" (37). If the bulk of commentators have regarded the Bartram book as "a work of art," others have classified it as a "work of science" (37). And although it is arguably the most well-established book of natural history in America prior to 1800, generic and evaluative disagreement still afflicts the *Travels*, eroding hope for consensus about lesser-known colonial works. (Irmscher steers clear of the nascent natural histories of earlier centuries, where harder distinctions remain to be made.) David Scofield Wilson separated out what he named "nature reportage" (7) from "natural history," while Irmscher more narrowly defined such reportage as "the genre in which free-lance, self-educated journeymen, only peripherally acquainted with the grand theories that justified their haphazard activities, wrote down their 'simple observations' meant to aid minds greater than theirs in the further refinement of their overarching systems and abstractions" (7). Irmscher's distinctions prove pertinent and true only after the Linnean taxonomy took shape in 1735. Before that watershed arrangement came into being, "systems and abstractions" of any sort were nonexistent, or honored more within antique Aristotelean observances than otherwise. American contributors to the Royal Society's *Transactions*, for instance, included the Reverend Cotton Mather, who found himself as welcome to address witchcraft and demonic possession in the Society's journal as plant hybridization. Ample discriminations for those scientific and literary works written during the pre-Bartram eras are lacking, which is where my chief interests lie.

Besides Irmscher's ambitious study, the most notable works came from Lee Alan Dugatkin, Pamela Regis, David Scofield Wilson, and Joseph Kastner. The Kastner book appeared in 1977. His "species of eternity" title alluded to eponymous plants named by naturalists: e.g., *Claytonia, Gardenia,* and *Franklinia*. Kastner was a former editor of *Life* magazine and surveyed the era dutifully, from Thomas Harriot in 1588 to the major work of John James Audubon in the 1840s, but within the first twelve of 350 pages he had arrived at the eighteenth century. Kastner also limited his book's value by omitting index, footnotes, or bibliography. He used "Bibliographic Notes"; and he chose not to load his fragmented quotations with ellipses because they "pockmarked the pages" (xii). Nonetheless, the Kastner book is valuable for its many insights and excellent prose. Wilson's 1978 book is scholarly. He found Bartram's *Travels* to be richer, not messier, for its fusion of genres, proving again that there is no accounting for taste in academe. Bartram "legitimized several options for expression of [his] energy—religion, botany, *belles lettres*," wrote Wilson (24), who offered full case studies of works by Jonathan Carver, John Bartram, and Mark Catesby. "The nature reporter was more than a scientist. He legitimated and domesticated as well as described American nature. Take cultural significance as the criterion instead of scientific worth, and a new picture emerges" (29). Wilson emphasized the eighteenth century exclusively. So did the 1992 book by Pamela Regis, who case-studied Bartram's *Travels*, Jefferson's *Notes on the State of Virginia*, and Crevecoeur's *Letters from an American Farmer*—all as "works of science" (xi). She placed those texts within the literatures of travel and natural history, defining the latter as "a broad area of scientific inquiry circumscribing the present-day disciplines of meteorology, geology, botany, zoology, and ethnology." Reviews of the Regis book were negative, or mixed at best. Most pertinent for ecocriticism is her argument that natural history initiated a new "literature of place" (5). Also, she wrote, "the Linnean method led to a literary style and to a rhetoric" that is singularly American (15). The most recent among these four books is by biologist

Lee Alan Dugatkin. Productively it strays (beyond Thomas Jefferson and the moose he used to try to disprove theories of American degeneracy in speciation) to argue for overlooked scientific thinking that was both ecological and paleontological. (Issued in 2009 by the University of Chicago Press, *Nature's Ghosts: Confronting Extinction from the Age of Jefferson to the Age of Ecology* by Mark V. Barrow, Jr., has a pertinent focus but a scope that regrettably excludes the seventeenth and eighteenth centuries.)

The limitations of all these scholarly treatments lie in their disciplinary orientation. The authors hail from the humanities, from the letters, and like most of us within the field they formally and critically treat their subjects. The early natural historians, though, were not men of letters; they were scientists, despite constraints imposed by the embryonic state of science at that time. One reviewer of the Regis book, in *Forest and Conservation History*, expressed his wish that "future work of this sort would be more balanced by a stronger appreciation of the keen grasp of material reality and the assiduously compiled detail of these accounts" (Stewart 142). Another of Regis's reviewers laments "the limitations of a literary study" as applied to natural history (Sheehan 255). Better sets of lenses might derive from the sciences themselves. David Barnhill developed one for analyzing nature writing. After comparing the taxonomies created independently by the ecocritics Thomas J. Lyon in 1989 and Patrick D. Murphy in 2000, Barnhill proposed an alternative that he dubbed "the ecosystem approach." His term *ecosystem* here suggests that the literary elements and the mixed genres intermesh in ways akin to natural systems. He argued that it "helps us to recognize the internal complexity of individual works by distinguishing ten basic elements found within texts." Those ten elements, reflecting the productive tension of the literary works themselves, entail "natural history, personal experience of nature, the social experience of nature, the philosophy of nature, an ecopsychology, the relation between nature and human language and knowledge, a philosophy of the human, ecosocial philosophy, praxis and politics, and spirituality." To appreciate how natural histories edge toward ecology, the ecosystem approach is apt, even if Barnhill's catalogue is a scattershot miscellany that includes the kitchen sink.

At a time when most literature was devotional or humanistic, early books about American flora and fauna offered alternative routes to the truth. They showed a healthy expansion of concerns beyond one's God or oneself, although humanism and mythology remained to complicate them. Mythology carried over from the Greeks and Romans, from Aristotle and Pliny especially, and from adherents in the Middle Ages. Christianity manifested chiefly in what we know now as the Argument from Design (Dugatkin 4). According to believers in that grand scheme, the Creation offered evidence of an omniscient Creator, and the manifest complexities of a natural order made that Creator all the more worthy of human worship and awe (this view relates closely to the strain of creationism known today as intelligent design). Humanism, which I use here to signify a philosophy that holds human culture and human technology as the apex of creation, also imbued our earliest natural histories. Accordingly the nascent nature writings most engaging and worthy of regard today are compounded, compromised, corrupted by utilitarianism—the ideology that nature holds value chiefly inasmuch as it addresses human needs for food, shelter, clothing, medicine, or profit. These works, due to their target readers and their purposes, commoditize nature. As a consequence some of the earliest natural histories are scarcely recognizable as such, so far do they diverge from the standards that have come to define natural history in the last two centuries. For these reasons alone, the scholarship is vexed and the assessments by their readers wildly divergent. Nonetheless, our earliest writings repay reflection and attention, partly for their continuity: for the ways their perplexing humanist and religious suppositions persist into the present day.

What best qualifies these writings as works of art, even if they are concurrently works of science, is an aspect of "enlightened humanism" that Robert Pogue Harrison has noted. Accordingly the North American forest is seen "as a consecrated place of oracular disclosures" (121) where an amazing rain of frogs might occur, as in a biblical revelation. The American forest may be "a place of strange or monstrous or enchanting epiphanies," like the black mass that Nathaniel Hawthorne's character Young Goodman Brown sees, or dreams, after which "A stern, a sad, a darkly meditative, a distrustful, if not a desperate, man did he become" (Hawthorne 95). Or, the forest may be "the imaginary site of lyric nostalgias and erotic errancy" (Harrison 121), as it was for Richard Foxwell, lost in a small boat at midnight, "fearing to land upon the barbarous shore," beholding "a great fire, and Men and Women hand in hand dancing round about it in a ring," who cried out his name mysteriously and urged him to join them in their frolics (Josselyn 21). Additionally, the forests could be "a natural sanctuary where wild animals may dwell in security far from the havoc of humanity" (Harrison 121), where the enmity shed on nature by human sin in Eden may be forgiven or forgotten (Isa. 11:6–8). The American forest was a site for new beginnings, a welcoming habitation, a place of open possibility.

A profile of Thomas Harriot imparts a sense of the privilege that cradled and cushioned many of the early "virtuosi"—the scientific specialists who left their written records of the continent's natural history. A graduate of Oxford, Harriot accompanied Sir Walter Raleigh to the Americas in 1585 as a partner in the venture at Roanoke Island. He learned and translated two Indian languages to assist the colony to settle. He practiced the craft of surveying, in which he was trained. After he returned to England, Harriot enjoyed the support of an early aficionado of science, Henry Percy, the Earl of Northumberland, who bestowed on Harriot an annual salary of £300, a sum that remained payable until his death. Harriot became one of the greatest scientists in the world, albeit one who did not receive due credit for his impacts on astronomy and mathematics. Harriot's sole gift to natural history, his *Briefe and True Report of the New Found Land of Virginia* (1588), remains an encyclopedic record of the species and people of that supposed New World.

No matter how prescient he proved in other sciences, Harriot eulogized tobacco. "It purgeth superfluous fleam [phlegm] and other grosse humours," Harriot wrote, and "openeth all the pores & passages of the body" (Achenbach). To open the pores should have meant dangerous and unwelcome sweating, but such was the state of science at the time. Nor was Harriot the only scientist to fall under tobacco's charms. After it was discovered in the Virginia Colony, it was universally thought to enhance human health. Its smoke was found useful for blisters, stomachaches, kidney stones, headaches, and even bad breath. The plant was a respected preventative of the plague. Boys at Eton College as late as eighteenth-century England were taught how to pack their pipes and inhale. They were required to carry pipes and tobacco in their bags. So serious were fears of the plague that boys might be whipped if they did not smoke regularly (Kastner 8). Now we know how mistaken such ecological interpretations of tobacco's impacts on the human body really are. Our grasp of medical science and human health has firmed up. Now we can ask how far such constructions might be attributable to bad science and how much to commerce. Maybe Harriot was willingly shilling for colonial tobacco farmers, covertly conducting a publicity campaign, a promotion entirely in keeping with the tobacco lobby today that still influences legislation. Maybe Harriot was simply susceptible to hearsay and lacking in empirical instincts. If beliefs about tobacco from 1588 can still ignite wonder, so do the writings outside English-speaking cultures in Harriot's era.

Bits of natural history inflect travel accounts by Columbus, Verrazzano, Las Casas, Cabeza de Vaca, Hernando de Soto, and Jean Ribaut. Utilitarian responses to nature dominated. French explorer Jacques Cartier, on reaching landfall at Funk Island off Newfoundland in 1534, stayed aboard while his "'men gathered two boatloads of great auk there, as Breton sailors relished their meat'" (qtd. in Morison 392). The quaint verb "gathered" indicates how easy it must have been to harvest that flightless bird, with its heft of five kilograms and wings only 15 centimeters in length (less than six inches). Native peoples had used it sustainably. A relative of the penguin, the great auk was extinct by 1844, though Cartier showed no inkling of extinction. The bird was at risk not only because it was flightless but also because it laid only a single egg. Robert Cavelier de La Salle marveled at the vast herds of "wild cattle," or bison, whose populations he found to be "almost beyond belief" (Kastner 9). Later writers, likewise mistaking the bison for a "cattle" or a "buffalo," showed how far they could assume the wooly beast's potential for domestication.

In 1635, English promoter William Wood likewise depicted moose in New England as useful oxlike creatures, so docile and kindly compliant that "the English have some thoughts of keeping them tame, and to accustome them to the yoake, which will be a great commoditie" (21). Anyone who has seen a bull moose hackle at close range, or a cow moose larger than a horse lay ears back and rush to protect her calf, will value the ludicrousness of such an assertion. Wood further finds cow moose exceedingly "uberous about the udder," or productive of milk (21), implying that these massive wild ungulates might consent to being made dairy animals as well as shackled to a plow. New England moose populations now are diminishing ("Researchers Track"), threatened perhaps by how much free meat they offer to the rapacious instincts that still afflict our species.

Were writers of some early American literature pulling the legs of their gullible English and European readers? Leslie A. Wardenaar thought so, and he analyzed the waggish appeals to nature in colonial nonfiction works. Folklorist Richard Dorson has likewise traced humorous exaggeration in these writings. From evidence in early natural histories, the American tall tale seems indeed to have its genesis in the seventeenth, not the nineteenth century as our conventional wisdom has reckoned. Subtle showmen of printed words, the earliest natural historians exploited the land's extravagant fables to lubricate colonial immigration. Christopher Levett, in *A Voyage into New England* (1628), claimed to have "omitted many things in this discourse" which would diminish his credibility, such as fish "with manes, eares and heads" like those on horses (n.p.). Levett's tantalizing anatomical specificity undercuts his disclaimer that he would prefer to omit potentially fantastic details than to itemize them and risk literary suspicion. The hitch inherent in classifying works like Levett's as promotional is that it is hard to feature how scary monsters could have spurred immigration. Our definitions of the genres need amending; what we have classed as promotional literature is often much more complex. A century and a half after Levett, William Bartram in his *Travels* (1791) must have sounded every bit as exotic and extravagant as the promoters when he issued a breathless description of the alligator: "The waters like a cataract descend from his opening jaws. Clouds of smoke issue from his dilated nostrils. The earth trembles with his thunder" (115). English audiences, reading rapt at hearthside, might have distinguished only with difficulty between these two accounts, the first of saltwater monsters in 1628, the second of freshwater monsters in 1791. Readers relished bizarre tales, as tabloids at our grocery checkout counters demonstrate some readers still do today, and accounts of America's natural plenitude, magnitude, and profusion afforded convenient literary recreation in the guise of probable fact. The problem of genre, as measured by current standards, still tips to the scientific. Readers could claim to be bettering themselves—learning about distant lands and creatures as in *National Geographic* today—while still getting vicarious thrills.

Some of the earliest natural historians wrote not only to attract potential colonists, but also to tantalize oversea patrons ravening for rarities. Patrons hungered for products they could trade on world markets, goods like "harts' horns" (i.e., deer antlers, still traded as restoratives or aphrodisiacs in Asia today) or wonders such as the Franklin tree, which wealthy private collectors including Peter Collinson coveted. Especially in the eighteenth century, specimen boxes routinely crossed the sea with contents of live birds, plants, and animals for European scrutiny and propagation. These were pursuits chiefly of well-born amateurs. Other naturalist-writers admired indigenous American singularities simply for the ways they added to the storehouse of human scientific knowledge, an impetus that became greater after the Royal Society, founded in 1660, began to issue its periodical. The no-man's land twixt fact and fiction was poorly lighted, as in Captain John Smith's 1625 account of cahow birds, a remarkable species which, "in the night if you but whoop and hollow, they will light upon you, that with your hands you may choose the fat and leave the leane" (2: 342). Smith's interests in natural history seem to have been chiefly gastronomic, an interest that was typical of many early travelers, promoters, and scientists. Settlers ate the cahow into near-extinction. (See the fuller discussion below.) Had other reporters, like Smith from the first three decades of the seventeenth century, not left records of this pelagic avian species, our knowledge of ecology would be poorer.

Even before the Enlightenment ushered in greater critical discernment, writers exhibited keen degrees of scientific disinterest. John Josselyn, for one, offered glimpses of ecological insight. His two books were published in the early 1670s. As late as 1759, Peter Collinson in England was recommending him to John Bartram (Irmscher 27). Evidently trained as a physician, Josselyn interspersed instances of *materia medica*—substances from nature to be deployed in medical remedies. Nor has pharmacology yet abandoned the forest for the laboratory. The mitotic inhibitor or cancer-fighter paclitaxel was isolated in recent years from the bark of the Pacific yew tree (Goodman and Walsh). If some of Josselyn's lore proves to be medieval in origin, other judgments he makes are empirically derived. He is remembered today not only in the Josselyn Botanical Society of Maine, the colony where he spent most of his time, but also for a standing as one who wrote "the most complete natural history of New England in the seventeenth century," in Donald Wharton's assertion (847). Besides the medical applications, Josselyn's books about New England catalog numerous now-defunct terms from natural history, such as *clicketing*, a word for wild canines that means to go into heat or copulate (59). In case Josselyn is mistaken for a mere creationist, a promoter of the Argument for Design like so many of his Puritan peers, we should note that he was fined three times—in 1665, 1667, and 1668—"for neglecting to attend church services in Maine" (*Two Voyages* xxii).

Josselyn gained authority by having sojourned twice—1638 to 1639, and again 1663 to 1671—and witnessing ecological changes over a thirty-three-year span of time, instead of relying on a single visit, as did many of his coevals. During his first voyage he reported flocks of passenger pigeons five miles long, but upon his second voyage he noted of the birds, entirely extirpated by 1914, "they are much diminished, the English taking them with nets" (*Two Voyages* 71). Clive Ponting placed the population of pigeons at that time at five billion (169) and noted that "their roosting areas were correspondingly enormous—some covered an area five miles by twelve with up to ninety nests in a single tree—branches broke and whole trees were toppled by the sheer weight of roosting birds, often standing on top of each other and leaving a pile of droppings several inches deep under the trees" (168). Enthusiasm for the new science of the eighteenth century has all but obscured the modest contributions of Josselyn in the seventeenth century, who could describe the North American marvel called the northern flying squirrel (*Glaucomys sabrinus)* in his *Two Voyages* with such perceptive analytical skill as this: "his skin being large and loose, he spreads it on both sides like wings when he passeth from one tree to another at great distance. I cannot call it flying nor leaping, for it is both" (62). Josselyn's work, such as it is, rewards reading and lays a sound foundation for Wharton's claim that he composed the most complete natural history of early America in his century.

By virtue of his long sojourns in New England, Josselyn brought profuse firsthand glimpses of nature reportage to the page. Form follows function in his sprawling sentence about a half-ton of wild flesh being chased, worn down, and slain by Native Americans in the Colony of Maine, a tale that smacks of a life-or-death quest he witnessed firsthand:

> when they light upon a Moose they run him down, which is sometimes in half a day, sometimes a whole day, but never give him over till they have tyred him, the snow being usually four foot deep, and the Beast very heavie he sinks every step, and as he runs sometimes bears down Arms of Trees that hang in his way, with his horns, as big as a mans thigh; other whiles, if any of their dogs (which are but small) come near, yerking out his heels (for he strikes like a horse) if a small Tree be in the way he breaks it quite asunder with one stroak, at last they get up to him on each side and transpierce him with their Lances, which formerly were no other but a staff of a yard and a half pointed with a Fishes bone made sharp at the end, but since [then] they put on pieces of sword-blades which they purchase of the *French* . . . they dart it into his sides, *haeret lateri lethalis arundo*; the poor Creature groans, and walks on heavily, for a space, then sinks and falls down like a ruined building, making the Earth to quake; then presently in come the Vic-

> tors, who having cut the throat of the slain take off his skin; their young webbs [wives] by this time are walking toward them with heavie bags and kettles at their backs, who laying down their burdens fall to work upon the Carkass, take out the heart, and from that the bone, cut off the left foot behind, take out the sinews, and cut out the tongue, &c. and as much of the Venison as will serve to satiate the hungry mawes of the Company: mean while the men pitch upon a place near some spring, and with their snow shoos shovel the snow away to the bare Earth in a circle, making round about a wall of snow; in the midst they make their *Vulcan* or fire near to a great Tree, upon the snags whereof they hang their kettles fill'd with the Venison; whilst that boils, the men after they have refresht themselves with a pipe of Tobacco dispose themselves to sleep (98).

Natural history and anthropology merge here, as Josselyn layers metaphor upon Virgilian metaphor and adds erudition to the forest scene. Dido, suffering for the love of Aeneas in the epic original, is likened to a wounded deer with an arrow in her side. The heart-throes of the anguished lover amplify the death-throes of the massive moose. Josselyn describes other mammals, all now evidently gone from the region, including woodland caribou and other species now indeterminate such the "*Mattrise,*" which is "a Creature whose head and fore parts is shaped somewhat like a Lyons, not altogether so big as a house-cat" (62). This creature, which he mysteriously characterizes elsewhere as a pack animal, might have been the marten or the fisher. To study species both extinct and extant, Capt. Smith's cahow and John Josselyn's moose alike, island ecology offers a useful science.

The biological dynamics of isolated islands, now known as island ecology, provide opportunities to study speciation and extinction patterns that generalize to larger ecosystems. At first limited to literal islands and the species richness they promote, island ecology now applies as well to ecosystems surrounded by mountains, water, fragmented forests, or by other natural or artificial geographical barriers or lines. On the matter of John Smith's cahow, which is a bird accepted today as a species of gadfly petrel, its range shrank following colonization of the western hemisphere, before it disappeared entirely from view. By 1901, Yale professor A. E. Verrill assumed it was extinct. Setbacks for this ground-nesting species, now identified as the Bermuda petrel (*Pterodroma cahow*), came not only from the colonists who captured and devoured it with ease, but especially from the pigs, rats, cats, and dogs the settlers released in Bermuda, effecting its extermination. Crosby dubbed such damaging dynamics "ecological imperialism." Garrard notes that "America's rapidly growing wealth in the nineteenth century was based on destruction and consumption of forests and wildlife so astonishingly voracious that, in places, it amounted to an 'ecocidal' campaign to exhaust and refashion whole habitats" (132-33). In fact, as the saga of the cahow shows, the ecocide began in the seventeenth century. In 1951, some fifty-one nesting cahow pairs were discovered on a single rocky islet off Bermuda. Now the bird, which is an ongoing project for restoration ecologists, has given rise to two documentary films and a book (Gehrman). Is the cahow a near-casualty of colonization? Or will its slow breeding, laying only one egg per year, take it out by natural selection? Theories of ecological sustainability and equilibrium can help address such questions.

Whereas ecologists of a generation ago liked to refer to *the balance of nature,* environmental ethicists and scientists now embrace disequilibrium as a theory that can qualify such older notions. Partridge analogizes it here: "The simplest example of an equilibrium would be a ball-bearing in a bowl. At rest, the ball is in the center. When jostled, it moves off-center, only to return to the exact same spot where it was before the disturbance. The shape of the bowl is the 'self-correcting mechanism.'" The bearing at rest in this analogy represents the balance of nature, the equilibrium. The "jostling" is the disruption of the natural balance or stasis that nature favors. By contrast, the theory of "disequilibrium" asserts that nature is in constant flux, or at best in states of provisional equilibrium, which of course is what evolution itself asserts. The idealized concept of a *climax community* of forest trees, for instance, is now called into question or overturned, depending on whom one consults. In literary studies, the closest equivalent to a climax ecological state is pastoralism, embodied well thematically by John Keats' "Ode on a Grecian Urn": "When old age shall this generation waste, / Thou shalt remain." Scientists and humanists today are debunking the idealistic stasis of pastoralism—drawn perhaps from the deep Christian well that promises eternal life in heaven. Ecocritics, in parallel with scientific disequilibrium, name the new paradigm "postequilibrium ecology" (Garrard 176) or "a postsustainability literary ecology" (Mentz 590). Biologists might rightly fret that such fluctuating worldviews and paradigm shifts will be seized upon to rationalize anthropogenic climate change and species extinctions. Another, competing notion driving ecological theory is "conservation reliance," developed by University of Idaho professor emeritus J. Michael Scott (Scott et al.; Mooallem 97, 114–15). So far has humankind disrupted ecosystem balances that some endangered species, those named the "conservation-reliant," will realize survival only if we regulate environs in their favor and intervene on their behalf. Is their threatened status evidence merely of disequilibrium? Hawaii's coastal plain wetlands alone are home to five endangered species of waterfowl that are being categorized as conservation-reliant and managed accordingly for survival.

The moose described in such detail by William Wood in 1635 and John Josselyn in 1674—that massive ungulate so iconic in New England culture, nearly seven feet high at the shoulder in some specimens—its habitat arguably has been "islanded." As Garrard noted, "Island extinctions have already been extensive; now habit destruction is also forcing mainland species into ever-diminishing ecosystems that are effectively 'islands'" (177). Civilization and sea circumscribe the wild environment of moose in New England, global climate change is now arguably permitting parasitic ticks to overwinter on its hide, and it is disappearing from its former range ("Researchers Track"). We may turn to natural histories of the colonial era as records of the preponderance and range of the moose, the patterns of reliance that sustained the Natives, the tools they used to hunt and kill it, and the "mattrise" that fed on its carcass in throngs. In short, colonial natural histories offer glimpses of interpretive ecology to be found nowhere else in the historical record.

One more set of pre-Enlightenment examples pertains. American Indians liked to leap astride swimming bears, or so Wood claimed, antedating the Davey Crockett legend in an image out of folklore motif indices. Such watery contests yielded "more sportfull Bear bayting than Paris Garden can affoard" (19). Such an appeal, tailored to European tastes for entertainment, hearkens back to that public recreation that delighted nobles including Queen Elizabeth. The amusement entailed staking a wild bear in an arena or public square and loosing hounds upon it (Elizabethan). Wood reports, too, that bears survive the New England winters by "sucking their pawes, which keepeth them as fat as they are in Summer" (20). This spurious account of hibernation traces to the Roman naturalist Pliny the Elder. Similar to ways that folk etymologies mistakenly identify the origins of words, so the folk science in Wood's imagery mistakes the fetal position bears adopt in hibernation. In their dens, scientists note today, bears may be seen "licking the bottom of their paws during the last half of the denning period when their old, callused footpads slough off" (National Park Service). The spirit of this homey image grows transcendental flesh and bones when it surfaces in *A Week on the Concord and Merrimack Rivers* where Thoreau adopts it as a trans-species emblem of self-reliance: "The poet is he that hath fat enough, like bears and marmots, to suck his claws all winter. He hibernates in this world, and feeds on his own marrow" (101). If Indians readily subdue bears in Wood's book, Josselyn cautions against too much complacency in the face of nature. Bears are highly unpredictable, during mating season especially, when "they walk the Country twenty, thirty, forty in a company, making a hideous noise with roaring, which you may hear a mile or two before they come so near to endanger the traveler" (*Rarities* 13). What might have sounded like three-dozen black bears were perhaps only two or three vocal bruins brawling or embroiled in combat, locked in sexual contests, their conflict witnessed only aurally and at a distance great enough for the cautious naturalist to craft his speculation.

These examples illustrate a key difference between natural histories of the early colonial period and those of the late colonial and national period. The earliest naturalists relied more readily on imagination, speculation, hyperbole, and secondary reports than on empirical observation, data collection, and firsthand knowledge. Many failed to heed the burden of proof that falls on people who seek to persuade, that is, the duty to offer evidence or cite their sources. They often resorted to verbal formulas (e.g., *It has been reported* or *A traveler held in high esteem has sworn*) in aiming to appraise the full range of American rarities. While Enlightenment intellectuals were inclined famously to "number the streaks of the tulip," as Samuel Johnson in *Rasselas* wrote, there is scant evidence, at least in the first decades of the Royal Society, of the modern scientific method that allows tentative truths to arise only from massive data. By the time Charles Willson Peale set up his Philadelphia Museum, the showmanship had evolved from the printed word to the portrait in paint, from the verbal rhetoric of praising American rarities and promoting immigration to the visual rhetoric of specimen collections and collectors.

Christoph Irmscher's *The Poetics of Natural History* has done for these vexed genres of environmental literature what Richard Dorson did for American folklore. While frontier naturalists like Josselyn reported species on the ground and in the raw, those collectors that most interest Irmscher paid for specimens. They yanked them from their native habitats to show them off to museumgoers and garden-strollers. They made their living in natural history. Ecologically speaking, the question that afflicts collectors of that era is whether the edification they brought to their visitors was ever repaid in species or habitat conservation, or whether in fact the collecting instinct decimated those same species. The Bartrams, Charles Willson Peale, P. T. Barnum, and others of their ilk practiced forms of natural history that would garner them good profits. They practiced arts that placed themselves on center stage. As Irmscher wrote, "American naturalists created texts in which they were the narrators as well as the spellbound protagonists" (4). Duped on some occasions, they created from those occasions a humble art that could dupe their audiences in turn, thus paying forward their gullibility in intergenerational succession. Likewise, as we see abundantly in works from prior centuries, "these reporters, instead of dutifully contributing to a fuller understanding of the myriad phenomena of nature, sometimes ended up perpetrating hoaxes instead of promoting the facts" (7). Remarkably, Irmscher here was speaking of more-sophisticated eighteenth- and nineteenth-century natural historians who were writing after Linnaeus came in. The challenging task for readers today is to distinguish hoax from artifice, fraud from artistic sleight of hand, particularly when it comes to the natural histories penned by seventeenth-century writers.

Among the annals of the collectors are some ecological moments worthy of note, thanks mostly to Irmscher's work on the subject. John and William Bartram both shipped seeds and plants to patrons in England, the elder to Peter Collinson chiefly, the son to John Fothergill. A unique flowering tree that enthralled the botanical community abroad, the Franklin tree, *Franklinia alatamaha* as William christened it, "occupies a special place as the ultimate collector's dream, a precious curiosity that ironically, almost from the moment it was discovered and gathered, had ceased to be available where it grew naturally" (Irmscher 42). Father and son found the tree in Georgia in Creek Indian territory. William named those Natives "'the generous and true sons of liberty'" (qtd. in Irmscher 63), alluding to the group of dissidents that undertook the Boston Tea Party in 1773. The species name for the tree, *alatamaha,* is of Timucua Indian origin and names a watershed that harbors some one hundred rare or endangered species. Another honorific note is evident in its genus, named for Benjamin Franklin, who in 1743 co-created the American Philosophical Society, the first learned society in the United States, and who as colonial Joint Postmaster General gave free postage to those naturalists to ship seeds and specimens abroad. The operative ecological question is whether William Bartram, who collected the Franklin tree and publicized it, is more to be praised for its preservation in his private garden or condemned for its destruction in its wild state. The enthralled international botanical community created a minor industry in "requests for specimens, especially from European collectors—a factor that might have contributed significantly to the complete removal of the plant from what was regarded as its native place" (Irmscher 50). Along with Irmscher we might conclude that, "Since it had ceased to exist in its natural habitat (whether this was its place of origin or not), the *Franklinia* eventually came to signify the inevitable damage done to the integrity of nature by the interference of the human observer" (51–52). What scientists call the observer effect is recognized as a factor in wildlife disturbance patterns today. Postcolonial naturalists, all ironically and unintentionally, might have loved this tree and other rare species to virtual extinction.

Early American collecting and displaying is shot through with ecological ironies. Collectors removed specimens from their ecological contexts, abstracted them in the form of interpretive notes or museum catalogues, and defined them in isolation from the webs of biological communities that gave them life and being. Museum "habitat groups" of the eighteenth century yielded to "dioramas" of the twentieth century. Those dioramas, life-sized displays that presented scenes from nature overlaid on painted or photographed backgrounds, attempted feebly to compensate for their ecological shortcomings. They gave the illusion of environmental reality via illuminating devices and other such forms of artifice. The genealogy of a pair of grizzly bears in the custody of Philadelphia artist and curator Charles Willson Peale exemplifies the problems. In 1807, the year after Lewis and Clark and their Corps of Discovery returned, Captain Zebulon Pike presented President Thomas Jefferson two grizzly bears as cubs. Pike and his men had bought them from Indians and transported them, "sometimes on their laps, for several thousand miles without significant problems" (Irmscher 76). Petted and kept inseparable, the bears were then displayed in Philadelphia confinement. As adults they became hopeless. One monkey, when it had the bad judgment to reach into the cage, had its arm torn off and eaten. Another monkey was dragged screaming through the bars by its chain and devoured. Irmscher sidesteps the denouement of this drama involving the bears, but suffice to say they ended as taxidermy models and were drawn by Peale's son Titian. Snatched from their ecological contexts, confined to conditions worse than a zoo, they became fated object lessons in collecting.

Peale and William Bartram, both residents of Philadelphia, were tightly allied. Peale painted what is now the standard portrait of Bartram in 1808, and Peale's son Titian retraced Bartram's travels to Florida in 1817–1818. Bartram committed himself to the place where he was born. The modest Quaker was a bookish homebody and a bioregionalist, by contemporary lights. Following a failed venture as a planter and four years of wandering to collect specimens and seeds, he settled at his home near Philadelphia and worked for 13 years on the manuscript of his only book, *Travels* (1891). Relished by Washington, Adams, and Jefferson at home, it went through dozens of editions abroad, influencing the verses of William Wordsworth and Samuel Taylor Coleridge. From its first paragraph, the *Travels* highlights the sublime powers of nature: "the wide ocean, which, a few moments past, was gentle and placid, is now thrown into disorder, and heaped into mountains, whose white curling crests seem to sweep the skies!" (29). A generation later, the Hudson River painters would adopt similar aesthetics. Rapture and effusion, features uncommon among the natural histories of its era, became Bartram's most compelling rhetorical forms, and for those he has suffered some unmerited loss of esteem in the last century at the hands of critics and historians, despite his shrewd contributions to ecology and his exacting sketches. Its reception reminds us there is little accounting for taste when a writer dares to break with received conventions, generic molds, and expectations. Enduring in its popularity, the Bartram book is also endearing in its exuberance, its persona so willing to articulate his rhapsody. In the years that followed its publication, Bartram declined to join the Corps of Discovery for which President Jefferson tried to muster him, declined to attend a meeting of the American Philosophical Society to which he had been elected, and even declined a professorship at the University of Pennsylvania.

As it did for Bartram, botany dominated natural history. Carl Linneaus of Sweden issued the first edition of his *Systema Naturae* in 1735, a book that revolutionized species identification and inspired a literary style and rhetoric (Regis 15). Botanists Mark Catesby, Alexander Garden, John Bartram, and his son William Bartram all integrated the Linnean method and sent plants to patrons in England who brokered them to others. Alexander Garden of Charleston complained that his desire to "botanize" distracted him from his work as a physician. Mark Catesby of

Carolina, a talented illustrator who also worked in botany, later known as "the colonial Audubon" for the quality of his drawings (Frick and Stearns), was the first to divine that birds of passage migrate to warmer climes instead of wintering underwater or in caves (Stearns 322).[1] William Bartram's father John, formally uneducated, identified 150 new botanical species, gained an international fame, and received a handsome salary once he became King's Botanist in 1765. Mosses of the genus *Bartramia* are the only plants that bear his name, but he nonetheless remains the first native-born naturalist of stature, in the estimation at least of Thomas Lyon (36).

One of John Bartram's closest allies in natural history, Thomas Jefferson, earned himself a place in the annals of science by virtue chiefly of his *Notes on the State of Virginia* (1787). Kastner called it "an admirably researched and constructed monograph on the natural history, climate, resources, and geography of his native state, along with dissertations on its education and politics" (121). Writing to William Bartram about some experiments he was conducting on flies for the American Philosophical Society, Jefferson confessed, "I long to be free for pursuits of this kind instead of the detestable ones in which I am now laboring" (120). Remarkable, Jefferson's statement, both as it assesses the drudgery of statecraft and the strong attraction of natural history for polymaths of Jefferson's time—even when the research was focusing on flies. Passion drove most of the natural historians and nature reporters. Jefferson's choice of naturalist Meriwether Lewis to lead the exploring party with William Clark across the American frontier proved to be an immense boon to the evolution of natural history and incipient ecology, despite his crude diction and syntax, as shown in this entry Clark penned upon the porcupine:

> the flesh of this anamal is a pleasant and whoalsome food—the quills had not yet obtained their usual length—it has four long toes, before on each foot, and the same number behind with the addition of one short one on each hind foot on the inner side. the toes of the feet are armed with long black nails particularly the fore feet—they weigh from 15 to 20 lbs—they resemble the slowth very much in the form of their hands, or fore feet. their teeth and eyes are like the bever.
> (Sept. 13, 1804)

Participating in a natural history dispute that divided nations, Jefferson was motivated in part to write his *Notes* to refute Buffon, de Pauw, and stodgy French *philosophes,* whom continental chauvinism had driven to assert that North American animals were only weak and degenerate relatives of their more powerful European cousins (Dugatkin ix). To counter the French naturalists, Jefferson quoted from Mark Catesby, Peter Kalm, and John Bartram.

To counter the French, Jefferson had a stuffed moose shipped from America to Europe as evidence of American supremacy, but he got in over his head when he waded into the debate about the mammoth remains discovered both in Russia and in Kentucky. As faulty as it sounds now, Jefferson held "that extinction is incommensurate with nature's laws" (Dugatkin 88). Such gaffes, ideological not ecological, subtract from his scientific credibility. So intent was he to combat the French, he overlooked the sound science propounded by Georges-Louis Leclerc, Count of Buffon, who "argued not only that the mammoth was extinct, but that extinction might be common" (85), a truism of evolution which presaged Darwin by almost a century, when Buffon made it in 1761. Indeed, "the whole debate around the bones of the mammoth and mastodon marked the birth of some of the tenets of modern paleontological thinking." Had scientists on both sides of the Atlantic had the foresight, they might have seen that forces anthropogenic as well as meteorological could impact both ecological relations and speciation. Jefferson's Deism could not admit of extinction and impeded the growth of his ecological thought.

The Philadelphia Quaker John Bartram found himself in hot theological water for his heretical views. His fellow worshippers disowned him for denying the divinity of Christ and for voicing opinions that make him sound like an advocate for animal rights today. "The creatures commonly called brutes," he wrote, "possess higher qualifications, and more exalted ideas, than our traditional mystery-mongers are willing to allow them" (Kastner 36). Animal welfare, animal rights, and the ethics that underlie them are at the cutting edge of ecocriticism today, even despite the antics of People for the Ethical Treatment of Animals, and Garrard in his widely adopted monograph devotes a chapter to the topic (146–80). John Bartram, a simple Friend, incurred the scorn of many competitive scientific virtuosi abroad for his lack of formal training. Jefferson surrounded himself with autodidacts of natural history who were gravitating toward the ecological.

From the standpoint of natural history, the most compelling examples from both centuries are those like William Bartram's that are "morally alive to the circumstances of the environmental moment" (Lyon 23). Quaker morality is notoriously progressive.

Let me conclude now by returning to my abiding interest and most singular focus of the present discussion: the neglected topic of interpretive ecology in early American literature. Surpassing John Josselyn of Maine and John Lawson of Carolina, both of them transplanted Englishmen, Peter Kalm, the Finnish-Swedish protégé of Linneaus, offered memorably incisive commentaries on American natural resources and on the ongoing impulse to exploit them unsustainably. He was "sent by the Swedish Academy to study the natural history of America" (Dugatkin 57). He spent two years in the colonies, 1748 and 1749, during which he successfully promulgated the reputation of Linneaus (Stearns 530). With Kalm's work, natural history made a decisive turn toward ecology. Arrived from Europe, where natural resources were already on the wane, Kalm warned against the same trends occurring in America. It distressed him to see colonists "bent only upon their present advantage, utterly regardless of posterity. By these means the swamps are already quite destitute of cedars" (Kalm 300). He decried Euroamericans' selfishly utilitarian attitudes toward the land and implicitly contrasted them with the legendary Native American belief in the need to consider the impacts of one's actions not only upon the present generation, but also upon seven generations into the future. Not all of Peter Kalm's observations were critical, though. He also "noted in passing that Catholic Canada was grateful to the Pope for classing the beaver as part fish so it could be eaten on fast days" (Kastner 37). As with the language that imbues so many seventeenth-century natural histories—those documents hybridized by promotion tracts, travelogues, tall tales, and paeans to God's glory—such statements might strike readers as slyly humorous today.

Kalm often combined observations of nature with social commentary. Perceiving an early need for hunting laws, he declared sagely, "In spring the people will steal eggs, mothers and young indifferently, because no regulations are made to the contrary. And if any had been made, the spirit of freedom which prevails in the country would not suffer them to be obeyed" (153). In the twenty-first century, a kindred spirit of deregulation countervails efforts toward preservation throughout the American West, the region of the nation that I know best. Reactionary voices loudly rise to endorse extractive practices on public lands, strengthen private property rights, and decry government efforts to enforce effective laws like the Endangered Species Act. As I write this, a standoff is taking place in Bunkerville, Nevada, where militias are assisting renegade rancher Cliven Bundy to hold off federal authorities that hope to force him to pay his grazing fees. Citizen militias are training their rifles on federal agents. Bundy's cattle meantime are encroaching illegally on critical habitat needed by the threatened desert tortoise. The "tragedy of the commons," long before Garrett Hardin brought the term into play, grew noticeable to travelers and first-generation immigrants like Crevecoeur, who lamented the effects of independent and antisocial behaviors. Hardin's phrase, adopted from economics, proves as pertinent within early American ecology as it is in the present day. The phrase alludes to the depletion of a shared resource by individuals who—acting alone, acting rationally in the scope of their private self-interest—understand that the depletion they are fomenting is contrary to the long-term comfort and security of others who enjoy the resource. In the case of the nesting birds in Kalm's example above, at least some of the plunderers must have been able to discern that their unsustainable spoils would propel vulnerable species toward extinction, like it did the dodo, the auk, and later the passenger pigeon.

Crevecoeur, the "American farmer" as he became known after the publication of his *Letters from an American Farmer*, espoused a liberal agrarian philosophy akin to what we know today as bioregionalism. Later labeled the "colonial Thoreau," Crevecoeur showed more interest in the pastoral dream of life within a decentralized society than in nature and ecology proper. Yet in his *Letters* he anticipated, by at least two centuries, contemporary America's natural-resource wars and interests in ecology, as when he noted old-growth forests were being logged unsustainably: "Our ancient woods kept the earth moist and damp, and the sun could evaporate none of the waters contained within their shades. Who knows how far these effects may extend?" (Lyon 35). Today's conservation biologists have set themselves the task of addressing such questions anew. Self-styled "combat scientists" (Wilkinson 2) allege in trepidation that cutting the last of the planet's ancient forests will hasten global climate change and the extinction of keystone species.

Scholars of literary history speak of the continuity of American poetry. Just so, the concerns of our earliest natural historians still occupy ecologists today. Hundreds of billions of taxpayer dollars are being spent every year to mitigate imperiled habitats, to recompense property owners for predation and takings, and to restore at-risk species—$10 billion alone thus far to restore salmon in the Columbia Basin and its tributaries (Settling Fish). There are few better ways to gain ecological perspective than to go back to the colonial and early national eras, when enlightened writers already were detecting problems that vex our environment today. If our interests are chiefly literary, we still may frame those early documents afresh by inflecting them ecologically. "Writing extinction involves not simply the problem of representing absence, but also the difficulty of narrating ongoing systemic crises within intrinsically individualizing forms such as the travelogue and the novel" (Garrard 178). Generically, Garrard suggested, we may view such texts afresh if we appreciate ecology as a shaping device, a determinant of forms.

In colonial nature writing, a factor of delight often embellishes the instruction. The world was so new, the species so abundant, and the potential for maintaining sublime wonder so great. Much of what we look for in natural histories nowadays can be found in early American literature and its hybrid nonfiction forms. Readers who are willing to take these antique scientists on their own terms can recover and appreciate their fine and affective prose, slight ecological insights, and sporadic notice to matters of sustainability. Despite the generic impurity of the documents they left behind, despite the dissension that complicates scholarly efforts to analyze and appreciate them, they hold up well beneath the scrutiny of all who believe the arts should edify, gratify, and activate.

NOTE

1. For all their shortcomings, early naturalists were not entirely incorrect in their surmises and assumptions about bird hibernation. The common poorwill has been shown to enter a state of torpor very similar to hibernation in order to survive long hot and cold spells.

TWO

Literary Activism and the Bioregional Agenda

Practitioners of the Foxfire approach to education say that students whose work gains purpose learn the most. Apprehending the products of scholastic endeavors makes a difference in the way the learning process itself proceeds. The proverbial paper chase, the appeal of good grades, these rarely prove satisfying goals in themselves. But factor in a practical consequence to those same academic exercises, add an element of instrumentality, and the learning game gains elevated stakes. Foxfire founder Eliot Wigginton, teaching high school in the Appalachian Mountains of northern Georgia in 1966, set out to invigorate his classrooms by dispatching his students into the community to gather information on folkways and folklore, crafts and arts, material culture and religious experience, all of which students collected into a quarterly magazine named *Foxfire* that was eventually printed in a series of illustrated commercial anthologies. As bioregionalist Kirkpatrick Sale has remarked of those early scholarly efforts: "though not every place has kept its history properly alive, a fountain of information still exists if we will but tap it—as shown, for example, in the wonderful Foxfire books, the recent collections of Indian lore, and many other projects of oral history and folk knowledge" (45). For those Foxfire students in Georgia, printing research and preserving knowledge gave their education meaning. Doubleday published the results of their labors in 1972 and sold millions. For other writers, the process and product of research and publication can be less fulfilling. As much as we might believe in the value of conducting scholarly dialogues, of advancing knowledge and offering new perspectives, a misgiving begins to gnaw—that most professors win grants, deliver papers, and write articles and books chiefly to gain tenure, promotion, and more grants. As much as we may underscore for student writers the value and necessity of the writing process, we still privilege the product in our writing careers more than we admit, still value destinations more than the journeys by which we get there. And even if we can agree that research and publication are inherently worthwhile, we may continue to wonder whether the bait of fame is luring us, say, or that elusive "disinterest" Matthew Arnold named the *sine qua non* of the highest scholarly criticism. Neither goal is sufficient in itself. Nor do the rewards of the tenure and promotion system always prove to be sufficient compensation. What's disparaged for decades as political correctness is in fact a sign that scholars are finding the relative inconsequentiality of much conventional research to be stifling and daunting. One key to getting beyond political correctness is to forge past narrow academic confines.

Bioregionalism may be broadly defined as the process of rediscovering human connections to the land. Although learning the lore implies the study of humanity's interpenetrations and interpretations of nature, Alexander Pope's injunction that "the proper study of mankind is man" must be rejected. The study of humankind cannot be conducted separately from the biological processes that nourish us, that allowed us to become the dominant species. All of us can walk the fields and forests of the particular regions we inhabit, learn to name flowers and recognize bird songs, gain a sense of how the snowbanks melt into the streams that become the river. The range of temperatures, the yearly precipitation, the creatures and species of the wild, the impacts of our own kind on the region—these details form the foundation of the sustainable culture that is the ultimate goal of bioregionalism. As it applies to the study of American literature, too, bioregionalism becomes a useful critical orientation insofar as it informs the mass of writing tied so intimately to nature, writing by people as varied as Anne Bradstreet and Gretel Ehrlich, Thomas Jefferson and Robinson Jeffers, Mark Twain and Willa Cather, not to mention early naturalists like John Josselyn, William Bartram, John James Audubon, and John Muir. Such an approach admittedly would have scarce bearing on the likes of Henry James, Edith Wharton, and many of the postmodernists, alienated by the natural world as they seem to be or, at the very least, separated from its embrace.

Sociologists tell us altruism ranks among the highest designs of humankind, beyond the pinnacle of the hierarchy of needs. If this is so, then reason tells us that scholars in the arts and letters need to conduct research that actually may effect change. Injustice is a menace even in the ivory tower. Yet we seem to suffer a rhetorical deficit, a vagueness of purpose. The most accomplished writing in other professions not only informs the unaware and prepares the uninitiated, not only renders an affective function and influences opinion: it also seeks to accomplish something hard and fast. It matters. Part of the anguish of John Berryman's poetic persona, Henry, arises from his belief that his works and days bear few consequences. The great literature he reads bores him, his friends bore him, "the tranquil hills, & gin look like a drag" (16). Once he publishes a book, he waits expectantly but can say nothing more about it than "no harm resulted from this" (82). Nor should it come as a surprise that unhappy Henry's author, feeling out of touch and ineffectual in his society, took his life in 1972, jumped from a bridge.

One of Berryman's younger peers, Robert Bly, prompts clever chuckles in some circles for his book *Iron John* and his work in the men's movement. Those chuckles demonstrate how writing meant to render certain social change is decried, twisted, misunderstood. (Foxfire founder Wigginton's name likewise has been blighted by the news of his seducing students.) Bly was a prominent poet and critic in the 1960s and 1970s, a practitioner of Deep Image poetry, one of a few canonical writers to imbue his work with staunch political opinions. Learning from the strategies of European and South American poets, he practiced a free verse that decried U.S. involvement in the Vietnam War. And in a series of essays, notably "Leaping Up into Political Poetry," he promoted a heightened awareness and a poetics of commitment that countered most canonized efforts in American letters. Bly argued that a "husk" had grown around the best impulses in the American psyche, and that those impulses require social outlets, but "the friction of the civil rights movement and the Vietnam War have worn the husk thin in a couple of places now" (246). A new swiftness and intensity is needed, Bly said, that would draw upon Whitman's critiques of the Civil War, Thoreau's objections to the Mexican War, and the scope of history that informs Ezra Pound's *Cantos*.

The politics of the 1960s charged American scholarship—as interest in matters of race, class, and gender still attest. Scholars have appropriated the tools of the social sciences and skillfully applied them in the letters and the arts. We still pay too much attention to scholarly audiences, though, attention that limits our rhetorical resources and stunts the written word's potential to engender much-needed change. All of us would benefit from more energy spent as public intellectuals. In "Trading on the Margin: Notes on the Culture of Criticism," Henry Louis Gates, Jr., commented, on the enervation possessing contemporary critics, "that it's mostly make-believe, that the brilliant Althusserian unmasking of the ideological apparatus of film editing you published in *October* won't even change the way Jon Peters or Mike Ovitz treats his secretary, let alone bring down the house of patriarchy" (181). And he is right. So much well-intended scholarly writing lacks a realistic purpose, lacks luster. It isn't reaching the readers who need most to hear and heed its message. Standard academic journals too often become the chief targets of learned discourse. Writers accordingly continue to preach to the converted and to overlook the most salient and essential questions. In these pages I propose a middle course between the demands of the academy and the needs of the American psyche as Bly defined them. In doing so I address not only scholarly writers but also poets, not only essayists but also novelists and playwrights, all of whom may influence those administrators charged with evaluating faculty performance.

ADVOCACY SCHOLARSHIP

The opening example of the Foxfire teaching method is not a gratuitous analogy. The deep engagement of Foxfire pupils, the student-centered learning, the student role in shaping curricula, and above all the practical consequences of the work itself serve as paradigm for what may be called a literary activism, a phrase that has come to mean all things to all people, but by which I intend a rhetoric of advocacy scholarship. In the case of early Foxfire students, the purpose was not only to inform and motivate readers but also to preserve ancient and traditional forms of knowledge, and thereby to weaken the powerful grasp of technology in contemporary life. It was a revolutionary agenda, a quiet one that practiced bioregionalism before the term came into currency.

Some of the great stigmas Whitman and Thoreau, Pound and Bly suffered in their time are sure to distort the ideas of scholars like Kirkpatrick Sale who sanction bioregionalism. Sale and others promote a rhetoric of advocacy, a literary activism. Not content merely to delineate his topic, Sale passionately advances the bioregional vision in books that range through sociology and economics, history and mythology, literature and science. Key to this philosophical agenda is "learning the lore" of the bioregions we inhabit, Sale insists, which forms a first step toward altering society as a whole. And yet interdisciplinary scholarly approaches, however highly touted, are more conspicuous in lip service than in practice, more observed in the breach than in actuality. If such allegiance were sincere, then bioregionalism would offer a tool to offset the trajectory of postmodernism on the one hand and the wide-eyed reverence of so much American nature writing on the other. From the bioregional vision, a muscular literary activism may result, one as rhetorically useful for the *belles lettres* as for scholarship, as pertinent to scholarly audiences as to the needs of government managers and the public at large.

The tools of the colonial historian and cultural critic are especially valuable in understanding bioregionalism. Imperialism—ecological as well as social—has been an ongoing pattern in America. In the same ways the Native Americans were crushed and overrun, wild species are reeling from the impacts of overharvesting, habitat eradication, and pollution. Native languages originally spoken in distinct regions, indigenous and immigrant languages alike, have been obliterated and the speakers forced to adapt to centralized and homogeneous standards; diverse forms of religion, a great cultural wealth, are stymied. As David McCloskey has noted, the latest phase is the decline of vitality of our city centers and hometowns, the depopulation of many rural areas, and a slow but steady urban sprawl that detaches people further from the earth (3–4). If the subjugation of nature involves the domination of people, subjugation of peoples correspondingly involves domination of nature. Flora, fauna, and people all suffer when a place is plundered. The resulting loss of diversity and vitality produces a double displacement, both natural and human, under an industrial global monoculture (7–8). Gary Snyder, a prominent spokesman for the movement, has defined bioregionalism as "the entry of place into the dialectic of history" (41), a definition that constitutes the poet's vote of confidence for the historian's long view, but one that also may be the basis for a rhetoric of scholarship we can call literary activism.

Ecologies and communities are two banks of the same stream. Having been lost together, they can also be restored together. Such a hope is central to the bioregional project of helping to seed new eco-cultures of peoples who know, love, and care for their lands. If the basic problem is displacement, then one answer is the reinhabitation of our bioregions. Again, with a poet's beguiling simplicity, Gary Snyder would remind us, "To know the spirit of a place is to realize that you are a part of a part and that the whole is made of parts, each of which is whole. You start with the part you are whole in" (38). Bioregionalism, then, aspires to replace arbitrary political jurisdictions with integrated native regions, to allow watersheds, ecoregions, and larger bioregions to become the basis of analysis, planning, and resource management. Scientifically, bioregionalism would join ecology to anthropology through geography by linking ecosystem, region, and culture. And from a sociological perspective, bioregionalism would generate a fresh social bond, one that would annul the ties of both ethnicity and bureaucracy. Politicians and political scientists like Frank Bryan and John McClaughry in Vermont have realized how efficiently decentralized state governments can function. Such a new regionalism does not advocate a return to tribalism, nor does it reinforce provincialism. Rather it calls instead for a new kind of ecocultural bond that is deeply rooted in the place itself.

My bioregion is known as the Palouse, a high area of rolling hills and fertile soil that encompasses two watersheds, one flowing west across the Columbia Plateau, the other south to the confluence of the Snake and Clearwater rivers, both ultimately draining into the Columbia River and west to the Pacific. The rainfall is slight, the snowpacks heavy, the winters long, the winds high, summers often sudden, dusty, and short. The bunchgrasses indigenous to the region, along with the lupine and paintbrush, penstemon and ponderosa pines, created rich diversity for species now mostly extirpated—gray wolf and grizzly bear, caribou and wolverine, sharptail grouse and prairie chicken. The native grasses have been all but supplanted by exotic invaders, human-introduced, including cheatgrass and knapweed in the early 1900s and yellow starthistle in the

1980s. Wheat and peas and lentils, crops on which the economy relies, blanket the hills. As timber prices rose, logging proceeded at a rapid pace; as more dams impeded the wild streams, salmon species faltered or failed; as farmers have relied increasingly on chemical inputs to maintain crop yields, the soil of the region has grown less stable and the waters less clear. But this litany of losses need not enervate the hopefulness of local activists; it need not overtop the optimistic and tenacious spirit of community. A tangible connectedness binds human inhabitants to one another and to the land, a bond that prompts a tacit satisfaction when bull moose wander into town, when wild onions and morel mushrooms come ripe, when Merriam's wild turkey populations burgeon and expand their range.

REGIONALISM AND BIOREGIONALISM

Tight connections ally bioregionalism and literary local color, known also as sectionalism or regionalism. American writers who identify closely with a geographical locale have historically endured the critical onus of provincialism, an onus grounded in insecurities that stem from comparisons to European culture. No one should forget how William Faulkner, irredeemably identified with rural Mississippi, had to struggle in relative obscurity for the first decades of his career. It took a Malcolm Cowley to persuade readers to look beyond Faulkner's regional setting to the universal themes that energize his work. Inherited wisdom still confers the belief that the chief features of regional literature include attention to the speech, manners, dress, dialect, and folkways of given geographical sections of the country. Yet readers who exercise due sensitivity to the finer dimensions of literary regionalism find that its focus on the more antique and outmoded forms of knowledge recommends it. Herbalism, today considered separately as the disciplines of botany and medicine, sprang from studying the species of a given region. Herbalism is one example of ways of knowing—often ignored by literary studies—that flourish in local-color fiction. Regionalism offers reminders that attending to "philosophy" in previous centuries meant reading all the physical sciences.

Approaching American literature bioregionally allows us to sift our indigenous cultures from our derivative cultures, our grassroots themes and motifs from the European inheritance, a task scarcely begun. Richard Dorson wrote surveys of early American lore—*Jonathan Draws the Longbow* and *America Begins*—but few scholars of history or literature see fit to reference these precious resources. Bioregionalism legitimizes such studies and investigates the ecological values and biological currents that run through so much American literature written prior to widespread urbanization. (Part of the bioregional project is to help urban dwellers learn that they rely on the countryside for many resources and may also act to help sustain rural resources.) To read American letters through varied strains of popular religion—prodigies, portents, supernatural wonders, myths, and lore—is to help level the social classes, to intermix the highbrow and the lowbrow, and to generate more comprehensive portraits of the periods. Just as essential for the bioregional approach, however, as novelist Thomas Hardy knew so well, is how mythic constructs reveal an intimate knowledge of place that owes no allegiance to the more conventional forms of wisdom inherited from other cultures and countries.

Historian David D. Hall identified a "lay tribalism" (156) that underlay the more salient and famous Puritanism, a tribalism that sprang from intimate affiliations with locale. The sickness of souls, the war Satan wages against Christ, the providence God offers righteous men and women, these are placeless themes. But the unique dynamics of Congregationalism in New England also lent themselves to the growth of "cunning folk," as they became widely known—indigenous herbalists and wizards ostracized from the dominant Christian circles, the hegemonic discourse forms. Young Goodman Brown of Nathaniel Hawthorne's famous tale undergoes his ordeal because the belief system of his community suggested it, because community culture had sown the seeds of his malaise. And everyone remembers the way that most memorable of literary cunning folk in America, Roger Chillingworth of *The Scarlet Letter*, uses his potent herbs and elixirs to poison the Reverend Arthur Dimmesdale. But scholars can look further than fiction for examples. The life of American diarist and midwife Martha Ballard, read and presented in a Pulitzer Prize-winning book by Laurel Thatcher Ulrich, shows how a cunning woman integrated with the early national folk cultures of New England.

Just as Anteus relied upon keeping his feet firmly planted on the earth to gain his power, so folk cultures remain most vital when their bioregional origins continue intact. The religious schisms that beset the seventeenth century may be understood in part as dictates imposed from outside the bioregions, exotic dogmas of continental origin. Jonathan Edwards, preaching and writing in the more rational eighteenth century, understood the threat of lay tribalism and tried to arrest it by personifying a deity whose potential for wrath—overshadowing the powers of nature and place—would briefly succeed in recapturing the sway that lay tribalism had come so near usurping. Never mind how sensational these early spiritual quests may sound to modern sensibilities now; they are the stuff of which our most vivid early literature is made. Sarah Orne Jewett's solitary souls in rural Maine owe much to their indigenous literary ancestry. Brockden Brown, Poe, and Melville created psychic dramas whose ties to bioregional roots have been unrecognized or neglected.

In a deceptively simple book that includes the essay "The Regional Motive," Wendell Berry explored bioregionalism before the term came into being. The Kentucky farmer and poet, writing in the wake of a supposedly reactionary agrarianism, may be said to have first given bioregionalism shape as it is today. Berry found *regionalism* in 1972 "a term [that] very quickly becomes an embarrassment or an obstruction." More than just diminutive in usage, the word for Wendell Berry could carry the negative freight of both nationalism—insular and patriotic—and of condescension, "which specializes in the quaint and the picturesque, and which behaves in general like an exploitive industry" (63). This is an especially relevant comparison given bioregionalism's intimate relationship with the environmental movement today. Berry's environmental bent is evident throughout but is compounded with an insistence that "local life is intricately dependent, for its quality but also for its continuance, upon local knowledge" (67). "The Regional Motive" accordingly celebrates the correspondence of human beings with biological processes, and it would remedy the sad fact that "the history of the white man's life on this continent has been to an alarming extent the history of the waste and exhaustion and degradation of the land" (65). To moderate this historical trajectory, bioregionalism explores numerous routes to sustainability—routes, that is, to sustain both agriculture and human culture. Less a theoretical construct imposed from afar than a delineation of human cultures as they clash or mesh with biological cultures, bioregionalism seeks to perpetuate the natural processes on which our economies rely, and to reach a steady-state cultural balance that can likewise be sustained. This breed of ecological rhetoric would use bioregionalism as the vehicle to drive toward the key goals of establishing a useful cultural criticism, reinvigorating the merit of scholarship, and practicing an ethics of responsible and effective activism.

ACTING OUT

Writers of the fine arts increasingly have put their talents to work toward activist ends. Gary Snyder's reading tours in the 1980s sometimes perplexed his audiences who expected poetry but instead got lectures that promoted bioregional principles. Terry Tempest Williams, a scientist by training, turned her talents toward emotional writings and readings that explore alternatives to human dominations of the land. Wallace Stegner and Edward Abbey crafted novels and essays that remain as political as they are poetical in advocating the preservation of the wild. These writers all write from the West, a measure not only of the heated battles over environmental issues but also of the importance of wild spaces to the western imagination. Insofar as writing constitutes an activism when it is wielded to effect political change, evidence of literary activism in the West is rife. Not content merely to inform or influence, the bioregional writer derives creative energy from what is widely known as "*praxis*, the union of philosophy and activism" (Scarce 32), a union not limited to radical actions. Such a union is essential. If philosophy that lacks a complementary impulse toward activism characterizes the most impotent scholarly writing in the arts and letters, then the impulse toward activism without a basis in concrete philosophy likewise can degenerate into anarchy or polemic.

Initiating a large departure in American nature writing, Rick Bass, author of the books *Oil Notes* and *The Ninemile Wolves* early in his career, *The Wild Marsh* and *The Black Rhinos of Namibia* more recently, set aside a career as a petroleum engineer and moved to Montana to devote himself to writing on nature and the environment. In the essay "20515 House, 20510 Senate," he articulated both the literary tensions and the activist passions that are combining in so many American writers. The title of Bass's essay offers a mnemonic for folks who want to know the zip codes for the addresses of Congress. Bass first characterizes the conflicting demands of politics and art, then hits upon a metaphor that crystallizes the key issues: "Suppose you are given a bucket of water. You're standing there holding it. Your home's on fire. Will you pour the cool water over the flames or will you sit there and write a poem about it?" Affective and instrumental ends conflict; art pulls the writer one way, practical duty another. The bucket of water may be construed as the writer's talent; the house afire is environmental degradation; the courses of action divide between politics and art. Bass follows with a passionate plea to lovers of nature writing to take the steps necessary to save Montana's Yaak Valley bioregion—"a sanctuary, a harbor, for wild things, and for diversity" (4)—and in so doing to assure that nature writers continue to enjoy the bioregions on which the art relies.

Just as a bucket of water is a feeble way to thwart a house afire, a single letter to Congress may seem futile beside the efforts of Washington lobbyists paid by special interests. The momentum of national politics often appears too imposing; the writer or would-be activist feels ineffectual and defeated from the start. But talented writers and speakers and organizers need not stop at a letter to Congress or even a letters campaign. In 1906 Upton Sinclair spurred the federal government to found and pass the Pure Food and Drug Act when he investigated the American meat packing industry with his novel *The Jungle*. The values of literary activism, using bioregional approaches, include an ability to impact wide audiences with well-placed books or articles; the opportunity to practice truly interdisciplinary forms of writing and research, not only within disciplines but across the genres; and the inherent rewards of altruistic efforts that can bring about social change. The bioregional agenda does not insist that environmental activism is more worthy in itself than activism for human rights, say, but it does insist that the subjugation of nature always involves domination of people, and the subjugation of peoples involves the domination of nature.

Scholar-critics are well prepared to oppose spoken and written mandates of government officers who undercut key bioregional efforts. Documents generated by out-of-touch civil servants who are ineffectually safeguarding the public trust can be obtained and read using the same critical skills we use to read literature. Environmental impact statements, offers of timber sales on public lands, state water appropriations bills, bids on livestock grazing allotments, and proposed legislation of all sorts are publicly available and often subject to public appeal. The rhetoric of literary activism can achieve these ends. Erik Ryberg worked at the Ecology Center at the University of Montana; he received graduate training in English before becoming an activist and later a lawyer; he successfully appealed a number of grazing allotment renewals and insists that such documents are "easily deconstructed." Combining critical discernment and bioregional knowledge learned firsthand from knowing the land, literary activists can offer perspectives on key issues that involve the wellbeing of all living creatures.

Participation in the democracy is at stake. With political apathy high and voter turnouts low, civil servants welcome public participation. The most vocal players involved in the democratic process in recent decades, members of the cornucopian or wise use movement, often have the greatest economic interests at stake and therefore constitute a partisan point of view. Other active participants also represent absentee interests, investors from outside the state or region who are apt to care less for the land than do those who live on and rely on it. Many land managers themselves welcome public input and are sensitive to the concerns of environmentalists and bioregionalists—as the advent of the three-thousand-member Forest Service Employees for Environmental Ethics attests—yet their hands often are tied by organizational superiors and by a hidebound tradition of business as usual. The would-be appellant or activist need not be versed in matters scientific or political to succeed; language is the vehicle by which most governmental officials conduct business, and language is subject to scrutiny and interpretation. Hidden motives of the writer(s), undue solicitude to special interests, lapses and lacunae in the arguments: the critical eye can discern such irregularities as these and refute them. Praxis, the union of philosophy and activism, can afford the authority and urgency required to bring about much-needed change.

In the brief case study here, I try to demonstrate these principles by rhetorically analyzing a document from the Forest Service. To constitute a fully bioregional study, of course, my discussion also would have to address the varied needs of Native Americans to maintain their cultural continuity and spirituality that rely upon intact ecosystems and bioregions, a task that I am undertaking now through conversations with the Upper Columbia United Tribes. Again, subjugation of nature always involves the subjugation of peoples.

GRAZING ON THE NORTH CASCADES

Horseshoe Basin is a high meadow on the extreme east side of the Pasayten Wilderness, part of the Okanogan National Forest (ONF), a popular recreation area in the North Cascades of Washington state. Legends tell of stirrup-high bunchgrasses that once flourished here and throughout the Pasayten, grasses now supplanted by ankle-high sod grasses and other weedy species. Water filtration has been severely curtailed, the dirt pounded for decade by thousands of hooves. Each summer some 1,200 sheep feed through the area as part of a grazing allotment. Associated ecological impacts of livestock grazing in the North Cascades—impacts surely not unique to the region—include high elevations and correspondingly short growing seasons, the close proximity of mining claims and grazing allotments, and the problem of hungry livestock grazing on "erosion control seedings," that is, on trees and grasses that are planted at taxpayer expense to diminish other ecological damages incurred by grazing, logging, and hiking.

In the checkerboard of state lands, federal lands, parks, and private inholdings that comprise the North Cascades, the ONF attempts to support about seventy-five separate grazing allotments, some founded before the turn of the century, many around the time of the Great Depression, others as late as 1973. These seventy-five allotments are forced to carry the weight of more than 14,000 animals, of which 11,756 are cattle. Range elevations vary from 2,500 to as high as 7,000 feet, and ONF maps demonstrate how very much of the forest is covered by livestock during the grazing season, usually June 1 to October 15, and how many allotments directly abut one another, even in fragile alpine sections. Short growing seasons, combined with relatively long grazing seasons, pose a grave danger to the overall health of the bioregion.

Area management plans reveal the high priority given to livestock interests. Passages in the unpublished and unpaginated ONF grazing report show how land managers often defer to a vocal constituency of cattlemen and miners. The agency "can not legally authorize grazing" on patented mining claims, it claims, for these claims are "classed private land." The report also notes, without any evidence of irony, that inactive mining claims above the Methow Valley "are situated on the steep south slope," and thus "do not have any significant impact on livestock grazing," as if mining claims might infringe on grazing privileges, whether or not such claims were on slopes too steep for stock to travel. The language reminds us that the acronym DNS, standing for "demonstration of nonsignificance," may be used to appease a variety of commodity-based interest groups competing for public-lands resources.

But commodity-based uses of the bioregion are not always competitive and antagonistic, as the following passage shows: "Logging roads and skid trails have made additional feed areas accessible for grazing."

That is, access roads and trails constructed to facilitate logging, inevitably damaging to wildlife populations and migrations, furnish forage sites for grazing cattle. What emerges from such observations is evidence of a sort of symbiotic relationship between logging and ranching. Loggers and grazing permit-holders alike may benefit from U.S. Forest Service roads that open up and scar the wilderness. Moreover, of course, logged tracts increase light available for growing plants and thus increase the forage available to cattle and sheep. In the North Cascades, as in much of the West, the Bureau of Land Management (BLM) and the U.S. Forest Service (USFS) often are forced to mediate between competing claims of livestock and logging interests, the most vocal champions of our much-abused public "land of multiple uses."

In a faint concession to critics of public grazing, ONF officials try to encourage grazing permit-holders to vary "turn-on locations," that is, switch the sites at which livestock owners release or "turn on" their animals to allotments each spring. Not only are herds most evident at these sites, but damage to stream-banks and vegetation is most conspicuous at turn-on locations where the animals bunch up. Resource management again becomes a game of appearances, played by means of ruses, like a beauty strip, a screen of trees left standing between a highway and a cutover forest, a trick that logging companies in the Pacific Northwest are learning with the hired help of landscape architects. If campers, hunters, and such "multiple-use beneficiaries" do not witness the damage sheep and cattle cause in wild areas, then for practical purposes no damage has occurred and Old West ways can continue unabated. Grazing seasons terminate just prior to the opening day of the state general hunting season, mid-October, a date that means cows must be out before hunters encounter them and grow suspicious that livestock are competing for forage with game animals. Most fauna in the West—deer, elk, black bear, and grouse, for instance—cannot thrive in livestock-larded habitat (Jacobs 292–305).

But the most telltale abuse of grazing privileges in the Okanogan bioregion is that scientists have been forced to serve cowboy functions. Grazing pressures on sensitive riparian zones in the ONF formerly were thwarted by state or federal employees "riding" herds of private livestock to disperse them—an absurd image and a scandalous use of public funds. Riding herds keeps them from bunching up and inflicting too much damage to the land. However, managers recently have advised permit-holders to begin "doing their own riding. The cattle need to be kept well dispersed especially out of plantations and out of riparian zones, off the main roads particularly in the fall when they bunch up against cattleguards." Again, appearances are foremost. The authors of the report do not detail what environmental damages might occur; merely, they advise owners to ride herds because other users of the public land might object to cattle defecating in streams and congesting busy roads. Other users in these multiple-use public lands travel roads for access, necessitating efforts to keep cattle off the roads where they will be readily seen.

The land seems to be developing mechanisms to defend itself. Grazing permit-holders, range-science specialists, and USFS employees alike have been chagrined to witness the proliferation and migration of poisonous and spiny varieties of weedy species like Russian and star thistle, halogeton, white top, tansy ragwort, and spotted knapweed. Some of these exotic plant species, which often compete with or prey on native plant species, also prove toxic to the domestic animals that help to spread them. Others, like the spotted knapweed, produce "allelopathic chemicals that inhibit the growth of other plants" (Rutledge and McLendon). One such chemical, cnicin, "inhibits root growth of other plants" and "destroys their ability to compete for limited soil moisture and nutrients. This allows spotted knapweed to form dense monocultures" (ibid.). The ecological irony is threefold: knapweed came to the United States in tainted seed used to sow forage for cattle, knapweed supplants native bunchgrasses and other forage on which cattle rely, and the disturbance regimes from the hooves of cattle on fragile public lands in the West create ideal conditions for knapweed to spread.

How do officials choose to deal with incursions of such plants? In heavily infested areas, the weeds are mowed and burned so that the seeds will not generate new plants. In other districts, herbicides are sprayed, much to the jeopardy of the bioregion at large. And in some districts, including parts of the North Cascades, noxious weeds are pulled by hand. The image is absurd: a government-hired laborer or scientist on public lands, bending over or falling to the knees to uproot weeds, his or her salary paid by tax dollars, his or her energy expended for the welfare of cattle ranchers and their herds first responsible for the spread of the weeds (Devine 44). U.S. ratepayers subsidize both the bargain-basement grazing rates and the salaries of civil servants laboring to defend them.

Sheep and cattle barons in the Old West fought range wars over pasture lands and waterways, real estate and market shares. But stakes in the New West have changed, and range wars now rage over issues like the preservation of native wildlife habitats and biological diversity, as the high-profile case of rancher Cliven Bundy in Nevada recently has shown (Brazile). The combatants include not only ranchers who graze their herds on public lands, sometimes like Bundy refusing to pay their grazing fees in defiance of federal laws on federal lands, but also critics armed with the latest data from conservation biology and economics. One of the largest unbroken wildernesses in the lower 48 states, the North Cascades warrants an immediate reprieve from cattle and sheep.

CONCLUSION

Academic writing demands a sense of consequentiality to give purpose to the work. Yet many educational practices in American schools isolate individual students and overemphasize learning by rote. Consequently, students gain few or no experiences in interdisciplinary approaches, enjoy little say in the direction of their studies, and find it difficult to relate their studies to life outside a classroom. Many of these same problems beset writers and researchers in colleges and universities. A lack of applicability characterizes a great deal of the work produced by professionals in the letters and arts. Writers yearn to prompt some practical consequences with their studies, yet such effects are rarely evident. Research projects often seem irrelevant to life beyond the professional conference or the library walls. The common saying "It's an academic question" demonstrates how many issues prove immaterial or unanswerable. The Foxfire method addresses this defect by assuring that the work is characterized by student action rather than by the passive reception of processed information. Literary activism, by any other name, possesses the same potential for adult writers, researchers, and ecocritics.

The most rudimentary knowledge of bioregionalism can provoke fresh perspectives in the disciplines. As a complement to ecocriticism, it offers scholars in the humanities a vital approach, one that would tie the humanities back to the land before they float off into an academic void. The greening of America, requisite as it is to the survival of species and culture alike, shows no sign of diminishing in significance for either urban or rural dwellers. And although space may be the final frontier, social and biological ecologies now are preoccupying many of our historians and artists, histories and art. Interdisciplinary approaches in scholarship, which encourage a merging of the sciences and arts, can prompt collaboration among researchers in disparate fields. But bioregionalism is more than mere philosophy or pedantry, more than some theory nice to contemplate. It truly gets us where we live. It teaches us the means by which we can reinvestigate our living regions and reinvest our selves.

THREE

West of Winthrop

Landscape and Language in Washington Territory

He was born in New Haven, Connecticut, twenty-four years before his tour of the West commenced. He traced a family line that stemmed back on his father's side to John Winthrop, first governor of the Massachusetts Bay. On his mother's side, his forebears include the evangelical Jonathan Edwards, the foremost theologian-philosopher of his time, and Timothy Dwight, of the literary coterie at Yale College known as the Connecticut Wits. Theodore Winthrop accordingly had the expectations of literary greatness thrust upon him. A scion of New England gentry, descendant of seven college presidents, he graduated atop his small Yale class in 1848. Thereafter he lived abroad, tried his hand at legal practices in St. Louis and New York, socialized with prominent landscape painters, and composed the four novels and three nonfiction books on which his slender reputation rests. He died young, in a botched Civil War battle he helped plan.

Winthrop's temper, education, and pedigree center him squarely in the master narratives of Euroamerica. Carrying money in his pocket and admirable attire in his saddlebags, he toured the western territories of California, Oregon, Washington, and British Columbia in 1853. His resulting adventure books set in the mountains of the West reveal the social constructions of nature and culture that were the inheritance of well-heeled New England aesthetes. What his books fail to disclose about nature and culture in his times is even more significant. Washington Territory in particular, where he spent most of his sojourn in the West, was a place of racial and ecological turmoil at the time he was writing. It was moving from an "open" frontier to a "closed" frontier, one where white dominance would become a permanent state (Lape 13). Indians were dying from diseases and bullets and drink, their economies transformed by white contact, their ancestral lands and historic lore subject to usurpation and ridicule. Winthrop alludes to few of these historical tensions, even though he traveled 320 miles under the guidance of members of the Klallam and Klickitat people, whose tribal names he hoped to borrow to title his most memorable book. Nine years later, however, a Confederate bullet and the prerogatives of his editors at Ticknor and Fields in Boston eclipsed his hopeful plans.

The most memorable of his books, a nonfiction travel account entitled *The Canoe and the Saddle*, and its companion text, the western-genre novel *John Brent*—both of them published posthumously in 1862—share documentary value for readers who enjoy exploration and adventure stories set in the time he lived. Those books divulge a great deal about the ways that privilege and presumption shaped the American West. The nonfiction book has accorded Winthrop a reputation as an ecological prophet of the Pacific Northwest who, long before the first Earth Day in 1970, composed a moving elegy for the untouched wilderness of America that he believed was disappearing fast. Critics have praised the book for decades. In hindsight, though, such a status might seem spurious and undeserved. For students of American culture, for ecocritics in particular, Winthrop's racial and religious intolerance or bigotry complicates or taints his record on the environment and the landscapes he traveled and elegized. His elitist outlook engenders a highly partisan view. In much the way a so-called beauty strip of trees along a highway camouflages clear-cut logging scars on distant hills, so the splendid scenery of Winthrop's travel narrative screens off offensive sites of conquest and imperial power.

In *The Canoe and the Saddle,* that most enduring of his western books, Winthrop trains a painterly eye on the seashores, peoples, forests, and storms of present-day Washington State. By education and inclination, Winthrop was a writer of the Hudson River School, a movement that manifests itself in rhapsodic scenes that attend keenly to color and landscape space. The painter Frederic Edwin Church—whose vast canvas *The Heart of the Andes* has been called "America's first one-man, one-picture blockbuster" (Hughes)—mentored Winthrop, befriended him, and served to model and inspire Winthrop's work. Considered ideologically, though, Hudson River artists collapse American history of the period into simplistic dimensions and ignore the injuries of Manifest Destiny in their earnest valorization of landscape. Using a theoretical concept that ecocritics and philosophers know today as ecopornography, we see that Winthrop's privileges and predilections inclined him toward an exploitative regard for the land.

TO FIND ONE'S WAY ALONG THROUGH LIFE

> "It was my own fault that I looked for gold-bearing quartz, and so found it bogus and a delusion. What right have we to demand the noble from the ignoble!"
> —*John Brent* (8–9)

Winthrop's father, a merchant who owned a library of 2,000 volumes, died when Theodore was twelve. Even at that age, though, the son could recognize that the family tradition was one of "intellectuality, patriotic soldiering, devotion to cause, and constant connection to Puritanism" (Colby 18). Admitted to Yale at fifteen, he was expelled early in his sophomore year for breaking windows in younger students' rooms (Eliot 6). He returned the next year, more compliant and probably contrite, "somewhat fastidious in dress," as a Yale classmate wrote, "but decidedly aristocratic and a man of high principles" (9). Following his college graduation, the young Winthrop was privileged enough to embark on a tour that took him throughout Western Europe, accorded him the affectation of one who knew his languages, and heightened his taste for art and cuisine.

The influence of visual arts on his developing sensibilities was profound. In Paris he spent time with his college classmate Richard Morris Hunt, who was then studying architecture, and with Richard's brother William Morris Hunt, who was studying art. Both men went on to make enduring reputations for themselves in their fields. Meantime, in 1850, two years after finishing at Yale, Winthrop was mapping out a book on art. All that would emerge from such ambitions, however, was a lecture entitled "The Fine Arts in America," delivered in 1856 (Woolf xi). In New York, at the Tenth Street Studio Building where Winthrop rented workspace in the mid-1850s, the painters Frederic Edwin Church and Sanford R. Gifford likewise maintained studios. Winthrop "delighted to haunt the studio of his friend Church, the painter, and watch day by day the progress of his picture, *The Heart of the Andes*" (Curtis 10), which was later displayed in the Tenth Street Studio, where a skylight naturally illuminated the commercially triumphant painting. Viewers paid twenty-five cents to see it. Church, an entrepreneur, had arranged for the concurrent publication of pamphlets of this painting by the Reverend Louis LeGrand and by Winthrop—his first published work. The painter "received an additional tribute in the form of a two-piano 'march' in New York by the composer George William Warren" (Carr 14). In the many years of their relationship, Winthrop and Church took journeys to the Adirondacks and Maine, one account of which Winthrop wrote up at great length as "Life in the Open Air"; there he coyly named Church "Iglesias," Spanish for *churches*. On that Maine trip, which took the men paddling a canoe and hiking far up Mt. Ktaadn, Winthrop watched Church work and praised his technique. Winthrop might have flourished as a gentleman of leisure and independent wealth, and his mother might have supported him in that ambition, but instead he felt driven to make himself a fortune.

In Paris he met American millionaire William H. Aspinwall, who took him under wing and became his periodic employer from 1851 to 1854. Winthrop first tutored Aspinwall's son in New York in 1851 then headed back to Europe to find the pupil a place at a Swiss school. In the states again, he bought paintings for Aspinwall for his art gallery. At one point he paid "$300 for a Thomas Cole landscape" (Cantwell 119), although he mostly copied papers and received a clerk's salary. Eventually he moved to the Pacific Mail Steamship Company that Aspinwall founded. That position took him traveling again, this time to Panama, where he sold tickets, guarded supplies and gold going to and from the California mines, and suffered fevers in the tropical heat. The job did not advance him as he hoped. And so he quit it and made his way up the western seaboard, exploring coastal California, Oregon, Washington, and Vancouver Island, gathering material for the western books that would make his literary reputation, always training an eye on the main chance, always on the lookout for ways to gain a solid fortune. He contemplated investing variously in cattle, gold, coal, and real estate. At Fort Dalles, though, on the Columbia River, he contracted smallpox. During his three-week convalescence, he indulged his abiding homesickness and drafted poems to himself and letters to his beloved sisters and mother.

Bending homeward at last, he traced his way on horseback via the Oregon Trail to St. Louis, a journey that gave him fodder for *John Brent*. The sentimental narrator of that prototype of a Western novel, in language that presages its Yankee author's final sacrifice in the Civil War, reflects, "It is easier to die for a holy cause than to find one's way along through life" (151).

By all accounts Winthrop was never well. Spiritually and physically a valetudinarian, disabled by dyspepsia but emotionally in a constant state of flux as well, he fretted over his poor health. One biographer called Winthrop's condition a "delicately constructed conscience that had not yet come to terms with itself" (Woolf 11). Another named him a "semi-invalid" (Williams 229), an often-sickly man who managed nonetheless to sprint across the Washington Territory in record time. While traveling the Oregon Trail toward home and family in New York, still shaky from the smallpox, he reported in his journal he had acquired "a little laudanum [&] a cholera powder—some saleratus [or baking soda]" (29). He dosed himself with this concoction. The baking soda might have alleviated his indigestion or dyspepsia, but the laudanum—a narcotic that contains opium or its derivatives—must have been taken to address greater pains.

He struggled earnestly with his faith, like many of his Puritan forebears. During college "he experienced a religious conversion and spent so many hours praying in his room that his sisters feared he was losing his mind" (Cantwell 116). In later years, "nervous exhaustion, vague and unlocalized illnesses, stomach pains and recurring attacks of crushing despondency alternated with brief periods of feverish sociability and enthusiasm." When he confessed by letter to his mother that his Christianity was faltering, she grieved to her friends, and he became the subject of much gossip in New York, New Haven, and Boston. One distant relative, George Templeton Strong, sniffed in his journal that Theodore, by this time living in New York, "was always letting off gas about atheism, and propounding horrible paradoxes. He takes every opportunity of giving judgment on heaven and earth and all things visible and invisible, and generally puts their alleged Creator out of court in a summary way" (450). Winthrop's struggle with religion also seeps into the books. The narrator-protagonist of *John Brent* complains that "he wanted to make the nation's prayers; but the [Puritans] pronounced his prayers Paganism" (44). Winthrop's journal profiles a spiritual man as well, an overreacher who believed in a kind of interspecies evolution. We live in an "unfinished world and when next great convulsions come we shall give place to other beings more advanced than us [;] who knows what places we shall take" (16). His discarded Christianity was giving way to other beliefs, unorthodox but equally hierarchical.

By August 21, 1853, the twenty-four-year-old traveler was wrapping up his Northwest excursion. His greatest adventure lay before him: a 330-mile sprint the length of the Washington territory. At rustic Port Townsend, in northern Puget Sound, he hired a forty-foot dugout canoe and Klallam paddlers to carry him some one hundred miles to Fort Nisqually in the southern reaches of the sound near present-day Tacoma. First he bartered harshly with them; in the canoe, he seized their liquor to keep them in line; he even flourished a Colt six-shooter to forestall a mutiny. That night he fed his paddlers well and watched them as they slept. Arrived at Fort Nisqually late the next day, he paid off his Klallam guides, hired a Klickitat scout for the overland route, then purchased "pork and hard-tack" (50) and three "scrubby mustangs" (59) before hitting the Naches Trail, bound for Fort Dalles on the Columbia River 220 miles away. Atop the Cascades, the 4,800-foot Naches Pass remained to be scaled, a treacherous task. Intimacies with his Klickitat guide soon soured, and on the east side of the Cascade mountain range Winthrop found himself lost and wandering in a desert prairie near Yakima, spurned, ill, and alone. His uncivil behavior toward his "frowzy" Indian companion (79) led to his desertion, as research from source materials some fifty years later shows. Following a drenched and restless night spent in a thunderstorm, Winthrop passed out, toppled from his horse, perhaps relapsing from the smallpox that had laid him low some weeks before.

Throughout these exploits he donned the identity of a cultural minister, in fine shirts and breeches and tasseled hats, one who felt called to proffer enlightenment and to leaven heavy episodes of Native American "savage ire suddenly flashing forth" (39), a statement that is retrospectively ironic, when we learn his ire flashed forth more often than the Natives he met. During his more reflective interludes he praised nature's glories, waxed ecstatic in their fine light, when his black puritanical moods did not intrude.

Racial tensions were high when he visited the Washington Territory and took notes for his book. Six years earlier, in 1847, Cayuse Indians had murdered Marcus Whitman and family in Walla Walla, an act that hardened extant animosities. The Gadsden Purchase of 1853, which "extended southward the New Mexico and Arizona territories, had completed the modern boundary of the United States" (Matthiessen 133) and taken Indian lands. Also in 1853, the very year he made his Northwest visit, Congress spun off the Washington Territory from the Oregon Territory. President Franklin Pierce appointed as Territorial Governor Isaac Stevens and charged him with surveying an overland rail route from the Great Lakes to Puget Sound. Stevens undertook his charge with "240 soldiers, surveyors, engineers, and naturalists" (Cantwell 147), before yielding the Pacific Railroad Survey to the other "young Napoleon," George B. McClellan (Richards 21). As Superintendent of Indian Affairs, McClellan enjoyed the convenient power to clear Indians from the lands that the new settlers were perceived to need. On the East side of the Cascades, an Indian chief named Kamiakin, a memorable figure in Winthrop's nonfiction narrative, "led a coalition of interior tribes against the Americans in what became known as the Yakima War of 1855–1856" (Ruby and Brown 274). These skirmishes followed a series of meetings, conducted by Stevens on both sides of the Cascades, to soften fractious Indians for treaties.

In his treaty proceedings, Governor Stevens disingenuously adopted the cramped and inadequate Chinook Jargon to render treaty terms and answer questions, even though translators were available who spoke both English and the regional Native languages. Winthrop learned to speak this jargon, too. In the jargon he rendered much of the stagey dialogue that passes for Indian-white communication in *The Canoe and the Saddle*. He prided himself on his polyglot command of all tongues, ancient and modern alike, and compiled a glossary of the jargon, this "incoherent coagulation of words" which he appended to the book (12). His glossary first seems prescient, a wise anthropological touch, until one recalls glossaries in the colonial promotion tracts from which Winthrop in his devotion to his ancestry was probably taking cues. Chinook Jargon was the medium of commercial exchange in the Northwest. Intrepid travelers, he wrote, adventurers and investors alike, would have benefited from some knowledge of this bastard vernacular "of English, French,

Spanish, Chinook, Kallapooga, Haida, and other tongues civilized and savage."

Winthrop's guide on the overland route, an Indian whom he knew as Loolowcan, came to Winthrop at the behest of the man's father, Yakama chief Owhi. Extensive research leads us to conclude with A. J. Splawn that this man actually was Lo-kout (129). As one Winthrop biographer has acknowledged, "Most of the Indians of the Northwest refused to give the white man their names, fearing the anger of their tamanous [vengeful god] if they did so" (Probst 241n.). Five years after Owhi arranged for his son to scout on behalf of the Yankee traveler, the chief would be captured and shot by cavalry near Spokane, Washington. Another son, Qualchan, would be hanged for an alleged murder. But Lo-kout, who was Winthrop's guide and Qualchan's brother, would slip the noose and live another fifty years. He survived Euroamerican exploration, settlement, and war. Aged but unyielding, he gave an interview in 1906 that Winthrop scholars unaccountably have overlooked. There, Lo-kout declared how much he wished that he had murdered Theodore Winthrop when he had the chance. The reasons for Lo-kout's bitter memories will come into fuller focus as his and Winthrop's life stories unfold below, but it might suffice now simply to say that Winthrop considered his Indian guides much less picturesque and fit for genteel delectation than he did nature's glories. Behind the screens of his beautiful trees, he abused his guides and contributed to the closing of the frontier.

In 1862, almost ten years after his tour of the Pacific Northwest concluded, Winthrop's books based on his western travels appeared. Generically, *The Canoe and the Saddle* is a travel account, a novelized memoir that is reminiscent of Thomas Morton's hybrid saga of life among the Indians titled *New English Canaan* (1637). Such a comparison breaks down, however, when we remember that Morton was a true friend of the Indians, an ally with them against "Captain Shrimp" (Miles Standish) and the other Puritan militants in Plymouth in his time. Winthrop's ur-Western novel, *John Brent*, explores themes of natural nobility as they play out amid the migrants who were trekking the Rocky Mountain basins and ranges of the Oregon Trail. *John Brent* was accepted for publication if the author agreed to cut an anti-slavery sequence, but he refused to do so (Eliot 14). To modern eyes, that pioneering equestrian saga is as flawed by religious bigotry, especially anti-Mormonism, as *The Canoe and the Saddle* is by its ethnocentric deprecation of the Natives. Nonetheless, the nonfiction book has received abounding praise from regional historians and ecocritics, such as Timothy Egan and Alan Weltzien.

Although in life he had failed as a writer—all his books rejected by publishers, due in part to his religious and ethnic prejudices—his death on the Civil War battlefield conferred upon him a notoriety that got publishers busy bidding for the same manuscripts they first had spurned. (It helped to boost his repute that the *Atlantic Monthly* had in its pipeline two of his early military accounts, articles that it issued in the days and weeks succeeding his death.) Winthrop set his best work in the American West, capturing the imagination of his readers in the East, so that "fifty-five editions of his five books appeared between 1861 and 1876, and Winthrop was, for nearly a decade, one of the most popular of American writers" (Cantwell 160). Many parts of the West as Winthrop shaped and shaded it still prove fascinating today, but particularly so his behind-the-scenes role in the closing of the frontier.

DEFENSIVE SPLENDOR

> It is unphilosophical to suppose that a strong race, developing under the best, largest, and calmest conditions of nature, will not achieve a destiny. (*The Canoe and the Saddle* 130)

A majority opinion among art historians is that the most commercially successful landscape painters advanced Manifest Destiny (Novak; Kelly; Wilton and Barringer). A belief in Manifest Destiny meant it was manifest—evident beyond all politics and laws—that the new nation was intended to muscle its way across the continent to the Pacific. The intention belonged to God. The best evidence lay in His handiwork as it manifests in nature. Such a belief, of course, situated God in opposition to the Native Americans of the West who stymied Euroamerican expansion, an opposition that was consistent with millennial prophecies that had enjoyed currency since colonial days. "As Tocqueville noted, 'honest citizens' commented to him: 'This world here belongs to us. . . . God, in refusing the first inhabitants the capacity to become civilised, has destined them in advance to inevitable destruction. The true owners of this continent are those who know how to take advantage of its riches'" (Novak 145). In harmony with Tocqueville's report, Theodore Winthrop meditates at length upon his own New England ancestors, whom "the indiscriminate revenge of the savage" forced into forts, much like those of the Hudson's Bay Company on Puget Sound where he found succor and supplies (*C&S* 54). To celebrate the natural splendors of creation was not only to praise God, therefore, but also to add defensive splendor to His impenetrable will that suggested the Indians needed to read the proverbial writing on the wall and yield to Christian dominance.

The commercial interests of the most successful landscape painters, Cole and Turner, Church and Bierstadt, for example, made them "adventurers" in the early American sense of the word. They were venture capitalists, willing to invest in journeys to the hinterlands to create splendid canvases to sell. Church and Bierstadt remain the most pertinent here. Significantly in their landscapes, nature is almost always barren of humankind. That illusion must have helped their audiences believe that Native Americans had not lived in these ecosystems for millennia, that

Euroamericans were not already at work building roadbeds and railways, trapping furs and felling trees, and that God's handiwork was still intact. Indeed, humankind is rarely part of nature at all, pictorially speaking, in the wish fulfillment that generates such landscape art. Such subterfuge today still is possible in the West, because some 48 percent of the land is held in public trust. It is a place, as the 1964 Wilderness Act says, where "man himself is a visitor who does not remain." In the case of the Pacific Northwest of Theodore Winthrop, spruces and firs at a great rate already were being cut and limbed and shipped to San Francisco—a city booming with wealth from the gold fields—to fashion piers and pilings and fine new homes. None of this industrial development intrudes on his bucolic narrative, however. Winthrop, an adventurer in the same dual sense as the colonists and landscape painters, traveled to a brutal land and shaped it in ways that he hoped would sell and augment him.

Literary studies of the nineteenth-century American West have refined theoretical discourse on Manifest Destiny as well. Richard Slotkin's work has proven enduring enough in its implications that a 1996 television special on visual art, entitled *The Wilderness and the West,* featured him as a speaker, while in 2013 an installment of *Bill Moyers & Company* interviewed Slotkin on guns and violence following the Newton, Connecticut, school shootings. Slotkin gained a complementary update in the study by Kris Fresonke, in whose view Manifest Destiny effectively supplanted God as a prompt in works of adventure and exploration like Winthrop's. Another way of saying Manifest Destiny, Fresonke wisely notes, is "the American territorial imperative" (106). Taking her cues from the writings of Ralph Waldo Emerson, Fresonke discovers "buoyant and apparently counterintuitive accounts of the dazzling abundance of nature, no longer proof of Christianity for Emerson, but of American exceptionality" (97). The exceptionality of Mt. Rainier in Winthrop's narrative rushes to mind in this context. Serene and imperial, it anchors and oversees tumultuous events in the region. It affords continuity and stability amid great racial and environmental chaos. It transcends the wreckage going on below. Too long have we glorified Europe, Winthrop said, sounding indebted to or derivative of Emerson. "Let us, therefore, develop our own world. It has taken us two centuries to discover our proper West across the Mississippi" (*C&S* 47–8). A nature-based nativism was busily supplanting God, but to read the book of nature with accuracy still meant to learn to observe divine designs and correspondences, as Governor John Winthrop was doing back in 1630, and thereby firm up one's imperial certitude. Like the howling wilderness that Governor Winthrop found, the West was wild, but God had blessed this West. As in Emerson and Winthrop, it is interesting to observe in Margaret Fuller and Henry Thoreau, Susan Rosowski has written, "the seriousness with which they accepted that the West was the American Muse" (3). A complicated muse, by turns violent and nurturing, gorgeous and grim, the

West was an evident synecdoche for nature's possibilities—and America's.

Besides the insightful work of art critics and literary critics, ecological criticism offers some productive approaches to adventurers like Winthrop. Indeed, the genesis of ecocriticism may be found in the American West. In 1993 the Western Literature Association gave rise to the Association for the Study of Literature and Environment, a stronghold for ecocritics today, an organization that has conducted biennial conferences and issued its own periodical in the intervening decades. The loci of the organization are in the West as well—the University of Nevada-Reno and more recently the University of Idaho. Indeed, the concentration of ecocritics in the Pacific Northwest is significant, what with *Journal of Ecocriticism* cofounder Rebecca Raglon at the University of British Columbia campus in Vancouver and author Greg Garrard at the Okanagan campus of UBC. Western literary scholars long have noted environmental tones in conversations about the West. Insofar as it might help to dislodge Winthrop's rhetoric of adventure within the Washington Territory, an apt term I would bring to bear is "ecopornography."

It does not yet appear in any standard dictionaries, but the word enjoys currency in a number of scholarly communities. It produces dozens of hits on the World Wide Web. It yokes the adjective *eco-* with the noun *pornography* to decry the analogous exploitations of nature and the human body. "Some environmentalists have spoken of stylized, exquisite nature photography and films as a kind of 'ecoporn' because such works objectify nature for human aesthetic pleasure" (Hegarty). Ecoporn often gives fraudulent views of nature void of human presence. To define it by example, one might do worse than this quotation from a recent novel set in the West by Annie Proulx, an adoptive Westerner herself as a resident of Wyoming: "Calendars, especially the scenic types with their glowing views of a world without telephone lines, rusting cars or burger stands, enraged him" (14–15). It is deceptive—and, Proulx suggests, a source of rage for some—to tart up "glowing views" of an allegedly wild nature that is barren of the debris of industry, views that disguise the ills that industrial civilization has caused. Environmentalist José Knighton has written disparagingly of landscape photography as ecoporn for the ways it inspires tourism and thereby increases the damage that comes from trampling public lands. Even in high school curricula the term has currency. A good online lesson (now deleted from the Web) defined it as words or images "used to denote claims made by individuals or corporations who wish to present themselves as environmentally conscientious but who are in fact merely using the language of environmental friendliness to score political gains or to increase sales" (Zoeckler).

The semantic range of ecopornography is wide. Often it is synonymous with the concept of "greenwashing," another hybrid neologism, a portmanteau that suggests the white of innocence may be surreptitiously supplanted with the green of environmental stewardship. Individuals, corporations, and governments in kind cleanse themselves of culpability by laving their names with that hue. A sort of protective coloration ensues, a green concealment that makes it harder for detractors to single them out for castigation or blame. The first use of the word traces to the early 1970s when adman Gerry Mander coined it to describe some of the aggressive ways that advertising usurps the flag of environmental propriety. Since he took leave of that industry, Mander has gone on to produce some provocative research on television, technology, and the Indian nations. Some years after Mander coined the term *ecopornography,* in a paper presented at a conference, a scholar could be found using Kenneth Burke and other rhetoricians to focus on ecopornography as "the visual rhetoric of stakeholder publications in contemporary and corporate culture" (Durfee 2). It is a term that fills a vacant niche in American diction, and ecopornography shows every indication of being here to stay. To my knowledge, no one yet has applied the concept to arts and artifacts prior to the twentieth century.

There is a variation on the pristine theme within the ecopornographic. This variation gives a vision of nature pleasantly altered by human impacts—changed forever but still pleasing to the eye. In a classic example from the 1960s and 1970s, the U.S. Bureau of Reclamation commissioned painters to portray results of agency efforts to "reclaim" the arid West. These artists transformed dammed rivers into sites where opportunities for recreation and scenic beauty still abound. The dozens of paintings include pieces by Richard Diebenkorn and Norman Rockwell, the latter perhaps the most beloved of American painters in his time. Rockwell painted the Glen Canyon Dam, unquestionably the most widely reviled of all water-development projects in the West. Rockwell posed a family of Navajo Indians on a bluff over the dam, as if to show how power generation and wasteland "reclamation" could coexist with traditional uses. Their dog cringes. A bald eagle and a red-tailed hawk soar above the dam face. More than a century after Winthrop visited the West, fine arts had come again to enhance the designs of Manifest Destiny. This time the U.S. government hired the artists outright.

WEST OF WINTHROP

> "I have really grown quite tired of green grass, and well-kept lawns, and the shaved, beardless, effeminate look of my native country. This rough nature is masculine. It reminds me of the youth of the world. I like to be in the presence of strong forces."
> —*John Brent* (122)

The Norman Rockwell image of hawks and Indians aiming to coexist with one of the world's largest dams brings us back to Winthrop, who enjoys a reputation as an avatar of environmental philosophy in the West. William Dean Howells wrote in praise of Winthrop's western work:

> Of the three novels *John Brent* is easily the most interesting by reason of its vigorous narrative of adventures in the Far West, at that time a region still barely touched by fiction, and its magnificent hero, the black horse Don Fulano. That Winthrop's real talent looked forward in this direction rather than backward to Hawthorne appears still more clearly from *The Canoe and the Saddle*, a fresh, vivid, amusing, and truthful record of his own journey across the Cascade Mountains, and an established classic of the North-west. (68)

Contemporary readers might agree that *The Canoe and the Saddle* is "fresh, vivid, and amusing" while nonetheless attending to its shortcomings. Winthrop's 1913 editor of the book, John H. Williams, lathered his introduction and notes with twisted opinions. Williams dubbed the Chinook Jargon the "deadliest of stupidities" (xxv), for instance, and apologized for the many Indian place names in the book. Those names are apt to "inspire horror and disgust" in the civilized reader (10n). Williams's other apparatus is rife with shortsightedness, too, as when he calls the book "a picture of the great stage set by Nature for the drama of state-building" (vii). In this way he underscores the Manifest Destiny theme. Williams was writing for readers of a century ago. Robert Cantwell, a historian writing in 1972, praised Winthrop for his analysis of ways the Northwest environment shapes character. Cantwell's lavish praise goes on for pages. "The influence of the wilderness was not a moral force in itself," he wrote of Winthrop's prescience; "it was powerful for what it said of life elsewhere" (164). Finally, journalist Timothy Egan, winner of both a Pulitzer Prize and a National Book Award, lauded the Winthrop book and adopted its author as a kind of spirit guide on his literary tour through the Pacific Northwest in *The Good Rain*. "The idea that mountains, especially ones with such an evil reputation, could be . . . transcendent . . . was visionary" (40, ellipses Egan's). He hints that Winthrop was "influenced by Ralph Waldo Emerson," a good possibility when one considers Winthrop's debt to Hudson River and luminist aesthetics, those visual-arts movements that found their "analogue in American Transcendentalism" (Radaker 267). Winthrop has attained lasting appeal among historians and literary scholars alike.

If Hudson River School painters in Winthrop's time favored mountains—the Andes, the Catskills, Adirondacks, Rockies, and Sierras—Winthrop brings the mighty Cascades to the page. His lengthy narrative of a summer thunderstorm atop Naches Pass is a model piece of ecoporn, reminiscent of Thomas Cole and Albert Bierstadt in its atmospherics. "A gloomy purple storm lay over the Cascades, vaster than they" (152). Winthrop, with his "trio of mustangs" and his Indian companion, each weary from a long day's ride, watches dark clouds advance from the direction they had come. "Beside that envelope of storm hiding the west from floor to cope, there was only to be seen, now softened with dull violet haze, the large, rude region of my day's gallop—thirty miles of surging earth, seamed with frequent valleys of streams flowing eastward, where scanty belts of timber grew by the water-side" (253). The landscape was alive in ways that rose above military might and white dominion. Such a scale might have astounded some of his readers back East. There is no cover to be had; the cavalcade stands vulnerable to the storm. "Fitful bursts of weeping rain were now coming thicker, until control ceased, and the floods fell with no interval, borne on furiously, dashing against every upright object as great crushing wave-walls smite on walls of cliff by the sea-side" (254). This is the sublime, as the landscape painters beheld it and fed it back to their viewers—so vast, grand, and perilous that it must inspire awe and veneration. "There were sudden clefts, and ravines with long sweeping flanks, and chasms where a cloud mountain-side had fallen in, leaving a precipice all ragged and ruinous, ready itself to fall" (255). The sublime also carries connotations of landscapes that are harsh, antisocial, and imperiously masculine. Magnificent and threatening at once, the scene recalls the many accomplished paintings of storms and mountains in the century.

The wild salmon, icons of the Northwest that ran some sixteen million strong in 1853, gave Winthrop opportunity to explore themes of barbarism and civility, loathsomeness and beauty. The native people of his acquaintance "were oozier with [salmon] juices than I could wish of people I must touch and smell for a voyage of two days" (20), but he buys salmon from them nonetheless and transforms the fish into objects of esteem through his highly developed culinary arts. If the people do not warrant his respect, the fish caught and sold by the people do. On the Naches Trail, preparing supper, he gives a painterly perception of a salmon he bought from Indians and props up to roast: "The colors that are encased within a salmon, awaiting fire that they may bloom, came forth artistically. On the toasted surface brightened warm yellows, and ruddy orange; and delicate pinkness, softened with downy gray, suffused the separating flakes" (74). His cultivated taste and touch transform the fish: "how much better than feeding foul Indians it was to belong to me, who would treat his proportions with respect, feel the exquisiteness of his coloring, grill him delicately, and eat him daintily!" (84). Salmon are impressive creatures, he concedes, even if he has to purchase them from "a singularly fishy old gentleman, his wife an oleaginous hag, [and] an emotionless youth" (85). The salmon offer him the opportunity to refine the coarse and clashing particulars of his western experience.

Literally and aesthetically, Mt. Rainier looms over Winthrop's account. At 14,409 feet (4,392 meters), it is the most heavily glaciated peak in the contiguous United States. The natives knew it only as Tacoma, an appellation that Winthrop sanctioned. Today that is the name of Washington's third-largest city. "It was a giant mountain dome of snow," Winthrop wrote of his first glimpse of the peak from the Indian canoe, "swelling and seeming to fill the aerial spheres as its image displaced the blue deeps of tranquil water" (43-4). The daydreaming young traveler, suddenly viewing this image reflected in the calm of Puget Sound, wonders if he is hallucinating. His Indian paddlers are plying the water, and he desires to escape their presence. Mountains exert an "ennobling influence" over humankind (49), and the Yankee traveler seems to have welcomed the reminder of his own gentility in the presence of his Klallam guides. In the peak's shadow especially, he felt nobler and more at ease, where "the summer evening air enfolded me sweetly, and down from the cliffs and snowy mounds of Tacoma a cool breeze fell like the spray of a cascade" (74). As if chosen by God, he fantasizes that the mountain is shedding the grace of its breeze for him alone. Aesthetically the mountain shares both male and female qualities, in the serene time of year that he passed through. There is "feminine beauty in the cones, and more of masculine force and hardihood in the rough pyramids." (45). Imagery of this sort drives home the pornographic elements in Winthrop's depictions of the scenic nature of the Pacific Northwest. His breathless renditions of Mt. Rainier help to camouflage the violence being accomplished below in the genocide of Indians and the extermination of creatures like the Plains bison on which the Indians relied for survival.

Numbing himself to the presence of his Indian companions, Winthrop studies Mt. Rainier from many angles, systematically and in detail as painters do, from all four points of the compass while he circumnavigates it on his journey to Fort Dalles. Here is an exacting view of what would be named, after him, the Winthrop Glacier on Mt. Rainier:

> The blue haze so wavered and trembled into sunlight, and sunbeams shot glimmering over snowy brinks so like a constant avalanche, that I might doubt whether this movement and waver and glimmer, this blending of mist with noontide flame, were not a drifting smoke and cloud of yellow sulphurous vapor floating over some slowly chilling crater far down in the red crevices. (125)

Something in the peak ethereal and permanent, remote and Olympian gives Winthrop distance from his present circumstances, allowing him to transcend territorial history and his participation in it. His dozens of classical allusions deepen the impression that he sees himself as outside time, that he is saved from barbaric damnation by his fine erudition.

Too often Winthrop grates today, environmentally speaking. He praises roads and rails, despite the destruction they bring, noting that "the unenlightened" always have disparaged "the destruction that precedes reconstruction" (102). This language is as rich in implications for the Indians as it is for the delicate fabric of ecological relations when habitats are fragmented to build civilizations. The Indians need to be spiritually destroyed, as does the forest, before they can be renovated. Like the Hudson River painters who influenced him, his remarks reside outside of necessity, shuttering history, privileging the present moment and the private perceiver. Peter Matthiessen recalled the actual empire that Winthrop seizes every occasion to praise. "Just after mid-century were undertaken the War Department surveys for a Pacific railroad," prompting "the final frenzied assault upon the bison, not only by trappers, Indians, and settlers but by professional meat hunters like Buffalo Bill Cody" (148). No meat hunter himself, Winthrop shoots and wounds a black bear for sport, vying vigorously to "discontinue the days of such shaggy monster" (189). The West seemed to have changed Winthrop's behavior. Washington Irving in his tour of the West wrote, "I found my ravenous and sanguinary propensities daily growing stronger upon the prairies" (90). Savagery and civilization are constantly in contest in Winthrop. In several amusing and revealing paeans to pork, his foodstuff of choice and a staple of the trail, Winthrop commends the cooking fumes of settlers wafting across the West as an essence of national culture (86, 170–71, 259). Pork fumes are good. By contrast, "the rude, dangerous forces of nature" (174) stand ever ready to tempt a man of quality. The Hudson's Bay Company forts where he sought refuge served trappers and traders engaged in exporting furs to Europe—a fact he never mentions. The mighty Mt. Rainier, object of his constant meditation and veneration, was one of the sole sources of stability in a world so swiftly shifting in matters of ecology and ethnicity.

Winthrop appears most hopelessly regressive in his attitudes toward ethnicity and race. He blasts the Indians with unflattering adjectives. His guide, "low-browed Loolowcan" (86), more accurately "a half-insolent, half-indifferent, jargoning savage" (183), possesses a "superstitious soul" (78). Prospective visitors to the territory should benefit from Winthrop's experiences and sagacity, he imagines, in the future "when the attempt is made to manage Pagan savages" (79). In the first few paragraphs of his narrative, he explains his haste to cross the territory and meet his traveling companions. He hires Indian paddlers and asks them to leave immediately. When they need to sleep off some liquor, Winthrop gets angry. "I became wroth, and, advancing where the king of all this region lay, limp, stertorous, and futile, I kicked him liberally" (9). "Yes!" he reports in triumph. "I have kicked a king." In that place and time, such a *coup de pied* was counted as the "deadliest of insults to an Indian" (Splawn 457n.). Regardless of their cultural practices, Winthrop exults in his waggish impudence and makes a mock of the occasion. Of course he gets his way. The Indians pack up and transport him, just as he commands, even sticking with him in their canoe when he confiscates their liquor and trains his pistol on them. "Look down this muzzle," he addresses them or pretends to, "as I whisk it about and bring it to bear on each of you in turn" (27).

Great bravado—actual or feigned, an aesthetic accouterment meant for effect—lards his narrative, demonstrates his domineering ways, and lays another bit of foundation for classic western literature.

Overlooked research by regional historian A. J. Splawn helps measure Winthrop's involvement in the violent culture collisions that were closing the Northwest frontier. In 1906, Splawn found and interviewed the Indian guide who deserted Winthrop fifty-three years before. Asked if he was Loolowcan of that adventure, the man "quickly rose to his feet; with flashing eyes, he said, 'Yes, I was then Loolowcan, but changed my name during the war later'" (129). At that time he was living with his hanged brother's widow, whom he had taken as his wife, near where the Spokane River meets and empties into the Columbia River in Eastern Washington. At an earlier encounter with the white militia, when his brother died in 1858, "Lo-kout, with two bullet holes through his breast fainted. A volunteer, in passing, struck him with his gun stock, in the forehead, crushing in his skull and leaving him for dead" (81). When Splawn interviewed him at age seventy-nine, "His skull had a hole in it that would hold an egg. How he ever survived such an injury I do not know." Lo-kout told Splawn, "Seven bullets have passed through my body and you see my skull has been crushed" (126). His experiences with Winthrop had helped to make a warrior of him. He said of Winthrop, in terms that *The Canoe and the Saddle* confirms,

> I did not like the man's looks and said so, but was ordered to get ready and start. He soon began to get cross and the farther we went the worse he got, and the night we stayed at the white men's camp who were working on the road in the mountains, he kicked me with his boot as if I was a dog. When we arrived at Wenas Creek, where some of our people were camped, I refused to go farther; he drew his revolver and told me I had to go with him to The Dalles. I would have killed him only for my cousin and aunt. I have often thought of that man and regretted I did not kill him. He was me-satch-ee (mean). (130)

Acting moody, despising his guide, delivering a kick while he slept, and using his pistol to get his way: these details line up well with the Yankee traveler as he profiles himself in his own book. No wonder he needed to turn away from his behavior, from his Indian guides, and from the domination of them by his own kind to nature as a transcendent space that language gave him power to control and praise.

Reading *The Canoe and the Saddle* less for its environmental prophecy and sensitivity than for its contribution to cultural studies and ecocriticism calls into question Winthrop's reputation and brings him into fuller focus for readers of the twenty-first century. As a reporter on the nineteenth-century American West, he told the truth as he saw it through his set of privileged lenses. If the truth his book gives is partisan and partial, still "the story of a civilized man's solitary onslaught at barbarism cannot lose its interest" (6), he wrote, in language that his audiences would have approved. Regional pride has vaunted him to the top ranks of travelers and writers in the Pacific Northwest. "He predicted," as Timothy Egan observes, "that a regional style and outlook would evolve as the North Cascades were appreciated for their singular beauty" (75). If that outlook often borders on the ecopornographic, "screening off the actual violence perpetrated there" (Bergthaller),[1] that geographic beauty remains alive despite his brutalizing of nature and his Indian guides.

NOTE

1. My analysis of Winthrop's travel narrative is indebted to Hannes Bergthaller, of National Chung-Hsing University, Taichung, Taiwan, from whose unpublished paper this quotation comes, and who generously shared his insightful work with me.

FOUR

An Iconography of Sabotage

Len Norris, a cartoonist for the *Vancouver Sun*, drew a man and woman in a backwoods cabin kitchen. Through their open door a smiling matron with a cup in hand advanced upon them. The women wore Christian head-coverings. The one indoors stirred a soup and said, "Here's that Mrs. Strippoff again to borrow another cup of gunpowder." The caption alluded to the forms of sabotage enacted by the Sons of Freedom Doukhobors, the separatist Russian immigrants to Canada who created the largest experiment in communal living in North America ("Lost Childhood"; see Fig. 4.1).

Mrs. Strippoff, a caricature, gained identity by the noise of her name. Her name played on the patronymic suffix (compare Smirnoff) to note she was prone to strip off her clothes and make a public protest. Her Doukhobor clan aimed to sabotage the composure of small-town British Columbia with nudity—and with gunpowder if need be. The Royal Canadian Mounted Police, allegedly to compel such Strippoffs to educate their children in public schools, raided homes in Krestova in 1953, confiscated some 200 children, and tutored them in an old sanatorium in New Denver (Androsoff 237). Wards of the Province, the kids remained there for the next six years, permitted visits from their families only through a cyclone fence for one hour every two weeks. To avenge such rough justice, those same Strippoffs later planted bombs or set fire to schools, libraries, bridges, railways, courthouses, and transmission towers ("Lost Childhood"). They also sometimes set fire to their own money and homes to protest bourgeois materialism, kneeling in prayerful ecstasy before the consuming flames. Canada labeled them terrorists and offered rewards for their capture. Len Norris hardened stereotypes and helped shape Canadian attitudes toward the Russian immigrant Christians. His cartoonish Doukhobors' casual regard for their fellows' nudity and bomb-

Figure 4.1. "Here's that Mrs. Stripoff again to borrow another cup of gunpowder." Len Norris, ca. 1950, reprint courtesy of Stephen Norris, the artist's executor and son.

making, downplaying the government affronts they underwent, made Mrs. Strippoff just another nettlesome neighbor who borrows goods instead of buying them for herself.

This brief introduction to the iconography of North American sabotage traces a line of anarchist culture across recent centuries. Forces that burnished these iconographic images include the American Revolution, the Luddites of the early=nineteenth-century English Midlands, the labor organizers in America named the Industrial Workers of the World (IWW), the writer Edward Abbey, radical activists like Earth First! and the Earth Liberation Front, and the War on Terror prosecuted by President George W. Bush. The U.S. Constitution, in Amendment I of the Bill of Rights, affords the free-speech guarantees that American writers and activists hold dear. It reads in full, "Congress shall make no law respecting an establishment of religion, or prohibiting the free exercise thereof; or abridging the freedom of speech, or of the press; or the right of the people peaceably to assemble, and to petition the government for a redress of grievances." Despite such constitutional promises of protection, champions of economic sabotage in the United States (as in Canada) have suffered mightily for their beliefs.

When freedom of speech failed for these groups and their aggravation grew, their speech acts took the form of vivid iconography, pictures akin to the "image events" that Delicath and DeLuca have identified as integral to public protests. Instead of being designed for televisual media, though, instead of being dynamic moments or photo-ops that would be transitory, these icons proved more culturally enduring. Today they often take the Internet as a medium, one that is free of charge and (so far) neutral in content. Conflicts are still playing out, though, about degrees of net neutrality for the future.

Sabotage as a strategy or tactic is distinct from other species of violence. At a time when deadly terrorism is tearing the foundations of governments and religions across the globe, sabotage must be defined carefully and viewed in proper perspective. The *OED* defines the word (first recorded in 1910) as "the malicious damaging or destruction of an employer's property by workmen during a strike or the like; hence *gen.* any disabling damage deliberately inflicted, esp. that carried out clandestinely in order to disrupt the economic or military resources of an enemy." The first rule of ecological sabotage is that "nobody gets hurt," that saboteurs draw no human blood, or so Edward Abbey asserted in his 1990 novel *Hayduke Lives!* (110). Its prequel, *The Monkey Wrench Gang* (1975), never has been made into a film, Calahan suggested, because "one can spray as much human blood across the big screen as one likes, but don't ever dare depict the sabotage of machinery!" The sabotage of machines, he observed, "violates the most sacred American value: property." An unauthorized and partial adaptation of *The Monkey Wrench Gang* was released in 2013 as *Night Moves*—a year after the director came under litigation for his appropriation of plot elements that Edward Abbey originated in 1975.

Property in America, according to some commentators, has a value that exceeds human life. Using these distinctions as pivot points, one may functionally define sabotage as an act of force or violence against property or decorum. If today's saboteurs invariably aim to cause economic sabotage and inflict the utmost discomfort on governments, contractors, hunters, and the like, the English Luddites (1811) harbored motives that proved much more fundamental in the legendary hierarchy of human needs. One ethical question among many that must be broached in this context is whether a person is more justified in performing economic sabotage in some conditions than in others. If the health of other species or people is at stake, is the sabotage more justifiable? Conversely, is sabotage less justifiable when economic self-interest is the motive? Using the hierarchy of needs again, it could be argued that altruism as a form of self-actualization, including altruism on behalf of non-humans, is more justified than acting out of strict self-interest. The Luddites were guarding their safety, the security of their jobs, their home economies.

The novelist Thomas Pynchon wisely saw a "clear identification between the first Luddites and our own revolutionary origins" (41). The Boston Tea Party of 1773, during which 342 chests of British tea valued at more than £10,000 were dumped into the water, is the prime example of sabotage as a tactic. The breakaway of the colonies during the American Revolution, exemplifying a class struggle won, informed the Luddite uprising and clarifies how sabotage might fit comfortably within North

American culture of recent centuries, both in Canada and the United States. Indeed, the Sons of Liberty who tossed the tea into Boston Harbor must have inspired the Canadian Sons of Freedom, aka the Freedomites or *Svobodniki*, in name alone if not in revolutionary fervor. The American revolutionaries hanged in effigy Andrew Oliver, who was the public official responsible for enforcing the Stamp Act, then marched on his home, smashed his windows, trashed his furniture, and looted his wine cellar ("Sons"). They did the same to the mansion of Thomas Hutchinson, the Lieutenant Governor and Chief Justice of Massachusetts.

The Luddite uprising began in England in 1811 with the advent of steam looms beset by skilled male laborers in the textile industry, particularly in the lace and stocking trades. Children as young as four and five, along with women, "came to make up roughly four fifths of the textile labor force by 1833, a population both easier to exploit and cheaper to hire than adult men" (Sale 33). Mortality rates were high among these workers, life expectancy rates low. Employment levels rose and fell to follow changing fashion trends. Harsh legal penalties accompanied these conditions in Britain, including death for the theft of a five-shilling pair of boots (82). Lesser crimes were punishable by transportation to the Australian colonies in the nineteenth century and to the American colonies in the eighteenth century. At the same time, rising literacy rates were elevating laborers' awareness of the causes and effects of the American and French revolutions; American writer Thomas Paine's *The Rights of Man* sold some 200,000 copies in Britain in 1793, an unprecedented number. Skilled craftspeople were being transformed into mere attendants on machines—"a grueling process of deskilling, depersonalizing, demoralizing, and degrading" (200)—and the indignation and resistance rose accordingly. Their resistance took the form of systematic campaigns to halt the mechanization of the textile industry. The term "Luddite" as used today, although varying widely in its definition, typically denotes one who regards technology with skepticism and sometimes refuses to use it.

In the counties of Nottinghamshire, Yorkshire, Lancashire, Derbyshire, and Cheshire, roving bands broke into factories at night and smashed the frames of steam-driven looms. They carried knives and pistols and threatened those who guarded the machines. They set fires and stole rifles and food. Some of these rebel bands were at least 2,000 strong. They addressed pseudonymous warning letters to manufacturers, and posted bills and manifestoes in public places. Their hero and probably apocryphal leader was Ned Ludd. His name survived to become a publisher of insurrectionary materials in the American West in the 1980s and the apocryphal voice of an advice column in the *Earth First! Journal* named Dear Ned Ludd. The first of several written Luddite manifestoes in England indicted all who would "'gain riches by the misery of his Fellow Creatures'" (qtd. in Sale 74). Such language unlocks the option, pertinent here, that "Fellow Creatures" might be broadly enough defined to include the non-human. English nobility attended closely to the threats and manifestoes. After Luddite sabotage turned to murder, the government began to act. By May 12, 1812, some 14,400 troops were sent from London to quell the Luddite uprisings. Poet George Gordon, Lord Byron, newly ascended to his seat in the House of Lords, "made the Luddite cause the subject of his maiden speech" (68n.) and later shared with peers his sorrow over "men sacrificed to improvements in mechanisms" (89). In a letter to Thomas More in 1816, Byron enclosed "an amiable *chanson*" he titled "Song for the Luddites," whose first stanza alludes to the American rebels "o'er the sea."

> As the Liberty lads o'er the sea
> Bought their freedom, and cheaply, with blood,
> So we, boys, we
> Will *die* fighting, or *live* free,
> And down with all kings but King Ludd!

Byron's reputation as a lover and a rascal has all but eclipsed his revolutionary tones.

The preferred tool of the Luddites when they set about smashing the frames of machines was the sledgehammer or blacksmith's hammer. This hammer has the longest genealogy in the iconography of sabotage. The late David Noble, author of a book-length essay on Luddism entitled *Progress Without People* (he was also a fired MIT professor and curator at the Smithsonian Institution in Washington), included as a frontispiece to his book a picture of himself with "Enoch's Hammer, the only Luddite sledgehammer still in existence" (ii). Such hammers are the oldest of saboteurs' tools.

Luddism and its specific stripe of sabotage voiced the rhetoric of class struggle. The plebeian wooden shoe, the clog or sabot, thrown into machines historically to break them, lent its name to the activity and became the second and arguably the most widely recognized of several objects to represent such sabotage. The sabot is a fit icon, inasmuch as it was the common shoe of the working classes. (One tentative title for the seminal novel *The Monkey Wrench Gang* was "The Wooden Shoe Gang.") At stake for the English Luddites was the ability to unionize, or in period terms "form combinations," and thereby stand in solidarity against the bosses' tendencies to exploit workers. These common goals placed the Luddites, the IWW, and later eco-warriors on parallel paths. The English Luddites, like their heirs today the Neo-Luddites (Glendinning; Jones), shared "a deep distrust of technology and resistance to its promises" (Sale 16). Violent and reactionary measures enacted by Parliament in 1812, similar to measures proposed by the U.S. government to eradicate radical environmental activism and sabotage, also link these radical special-interest groups (see "Countering Eco-Terrorism" and Veasaw).

Later but structurally related were the Swing Riots of the 1830s in England, when arsonists and machine-wreckers reacted to the advent of threshing machines. Some 600 people were imprisoned, 500 transported, and dozens hanged or otherwise executed.

THE WOBBLIES

The Industrial Workers of the World originated in the United States in 1905, according to its website, "to confront the global power of the bosses." The group claimed 100,000 members. Colloquially known as the Wobs or Wobblies, the IWW has worked to organize people, not jobs, by "agitating around particular issues or grievances in a specific workplace, or across an industry." This affinity for issues and industries is yet another of the many ways that the Wobblies bear comparison to environmental radicals like Earth First!, whose acts of ecotage landed them atop U.S. domestic terrorist lists in recent decades. The Wobblies' use of humor and high spirits, their art and music, made their rabble-rousing projects seem like hijinks merely. But those early labor organizers were certainly earnest in their determination to work outside the political system and to use sabotage, as the need arose, to satisfy their most cherished goals. They borrowed direct-action tactics from the Luddites, from the Boston Tea Party, from nineteenth-century American abolitionists, from "Resistance to Civil Government" (aka "Civil Disobedience") by Henry David Thoreau, and from an epigram written by the American poet Walt Whitman: "Resist much, obey little." The Wobblies shared boldness, chutzpah, a refusal to compromise, and some would say a disposition to be martyrs.

Members of the IWW called themselves "reds," but a glance at the writings of Karl Marx sets them apart from the communists. Marx was aware of Luddites and their ordeal, and in *Kapital* he discussed the revolts of European workers against machinery and large-scale industry. Marx noted that capital had overcome resistance to the advent of its machines only with great difficulty. He wrote of the original Luddite uprising, "It took both time and experience before the workers learnt to distinguish between machinery and its employment by capital, and therefore to transfer their attacks from the material instruments of production to the form of society which utilizes these instruments" (554). As this statement and others show, Marx appeared to have little sympathy or patience for those who struck out at the machines or "instruments" themselves, rather than working to effect social change on a deeper level. Here again the distinction between economic sabotage and the toppling of capitalist leaders needs to be made. Marx and early European labor organizers seemed to welcome the advent of laborsaving machines that they supposed would spare the working classes from undue strain.

Like the members of Earth First! that would follow them some sixty-five years later, the Wobblies adopted the use of "silent agitators"—icons drawn or plastered at the factory or sawmill, on machines or on windows, to let the bosses know that potential saboteurs were watching, stealthy as cats. The black cat, a favorite icon, is one that the Wobblies knew as "sabo-tabby" (Fig 4.2). With its arched back, ominous eyes, raised tail, and open mouth, the cat supplanted human rationality with atavistic gut instinct in its readiness to strike out. Black cats also hold centuries of connotations that call up hazards and bad luck.

Both the Wobblies and environmentalists integrated songbooks—*The Little Red Song Book* for the Wobblies, the *Li'l Green Songbook* for the followers of Greenpeace and Earth First! Songs are only one of many tactics radical environmentalists learned from the Wobblies, whose purpose was to utilize music "to fan the flames of discontent," which was in fact the subtitle of the book. These cross-pollinated songbooks both inspired followers and converted visitors to the fun and the art that were part of the movement toward greater egalitarianism. Indeed, one reviewer has named the Wobblies "a singing movement" (Zisk). They made anarchy amusing and aerated unhealthy rage.

Songbooks in American culture began, innocuously enough, as collections used for singing carols at Christmas or for harmonizing around campfires. The Salvation Army, a religious relief society, had its own songbook, and the group clashed with the IWW at public demonstrations. According to Trombold, "competition with the Salvation Army led the IWW to develop its own songs as an organizing countermeasure" (301). Songs and songbooks for the Wobblies became rallying choruses and competitive devices, proving that "musical parody was an act of political appropriation" (302). If such songs as satirical mimicry were tools to be utilized for sabotage, they made the singers themselves "guerrilla minstrels," in the phrase of Hampton: versifiers who sometimes worked under pseudonyms and usually alone. Such songs were not confined to American anarchists, of course, as demonstrated by the epithet *chansons de geste proletaire* ("songs of workers' deeds") from the French, a nation whose language gave rise to the locution *sabotage*. A number of American eco-troubadours today focus on environmental topics, including Darryl Cherney, Alice DiMicele, Robert Hoyt, Dana Lyons, and Casey Neil.

Joe Hill, executed by firing squad in Utah in 1915, after he was convicted of murder or maybe framed for it, wrote many of the Wobblies' songs. He was an active organizer and member of the IWW. His works and person have attained iconographic status. Wallace Stegner wrote a biographical novel about him. He is best remembered as a songwriter, as a prototype of the protest music tradition that came to be perpetuated throughout the century by such minstrels as Woody Guthrie, Pete Seeger, Utah Phillips, Bob Dylan, Joan Baez, Patti Smith, and Ani DiFranco. Some

Figure 4.2. Ralph Chaplin (aka Bingo), Sabo-Tabby, ca. 1915, reprint courtesy of Industrial Workers of the World. ©Industrial Workers of the World Union

of Joe Hill's lyrics contain references to outright sabotage, while many others advocate sabotage in effect, whether by promoting class warfare, defying exploitative employers, or by regarding with suspicion the easy assurances of the Christian faith. The following lyrics are taken from Joe Hill's adaptation of the 1891 song "Ta-Ra-Ra-Boom-De-Ay":

> I had a job once threshing wheat, worked sixteen hours with hands and feet.
> And when the moon was shining bright, they kept me working all the night.
> One moonlight night, I hate to tell, I "accidentally" slipped and fell.

> My pitchfork went right in between some cog wheels of that thresh-machine.

Note the insistence upon the moonlight—a setting necessary for performing illegal night work (or as Edward Abbey gave it, *Nachtwerk* [qtd. in Zakin 9], also the name of the oldest dance hall in Munich, Germany). Note too the rural setting, a wheat farm, where the work can grow so hot from the harvest sun and itchy from the chaff of the wheat that settles in the worker's skin and hair. Of these lyrics, Gibbs M. Smith has written,

> Hill also created a song reminding the workers that a little "sabotage" was sometimes necessary to help an employer realize the logic of his demands.
>
> "Conscientious withdrawal of efficiency" was not a mandate for violence, but rather for a sprinkling of sand in the workings of a machine, mysteriously broken bands around bundles of shingles, or, perhaps, as Hill illustrated in "Ta-ra-ra Boom De-Ay," a slight slip at the right time. (35)

The chorus of this song continues, with a tone of jubilation that accompanies the prospect of much-needed rest, after the speaker has done a double shift of hard manual labor,

> Ta-ra-ra-boom-de-ay!
> It made a noise that way.
> And wheels and bolts and hay,
> Went flying every way.
> That stingy rube said, "Well!
> A thousand gone to hell."
> But I did sleep that night,
> I needed it all right.

The stingy "rube" of these lyrics—a hick or unsophisticated person from a rural area—is the boss who insists on the workers' enduring sixteen-hour shifts. By identifying the boss as unworldly in this fashion, the worker or speaker positions himself as worldlier and wiser by contrast. The worker knows how to assuage his dissatisfaction and get some needed rest by revenging himself upon the boss through sabotage. The lyrics continue:

> Next day that stingy rube did say, "I'll bring my eggs to town today;
> You grease my wagon up, you mutt, and don't forget to screw the nut."
> I greased his wagon all right, but I plumb forgot to screw the nut,
> And when he started on that trip, the wheel slipped off and broke his hip.

Here the lyrics ratchet up the language of class struggle. The boss calls the worker by a name usually reserved for crossed-bred dogs, "mutt." If the worker views the boss as unsophisticated, the employer names the worker a mongrel or hybrid animal, whose mixed blood makes him squalid or impure. The lyrics intensify the threat of violence, too, inasmuch as the worker's sabotage results in a broken hip for the boss. It's hard to agree with Smith that the song is "not a mandate for violence" when the outcome is so grim.

Ta-ra-ra-boom-de-ay!
It made a noise that way,
That rube was sure a sight,
And mad enough to fight;
His whiskers and his legs
Were full of scrambled eggs;
I told him, "That's too bad—
I'm feeling very sad."

The imagery here draws upon pratfalls of slapstick comedy. It reduces the broken hip and convalescence to the stock and jocular saying, "egg on the face." One who suffers egg on the face has not been cautious, has not heeded signs or listened to wise words of warning.

And then that farmer said, "You turk! I bet you are an I-Won't-Work."
He paid me off right there, By Gum! So I went home and told my chum.
Next day when threshing did commence, my chum was Johnny on the fence;
And 'pon my word, that awkward kid, he dropped his pitchfork, like I did.

The boss recognizes the worker as a member of the IWW (pejoratively, "I-Won't-Work"). Strikes, slow-downs, walkouts, idling, and machine wrecking: by these devices the Wobblies earned reputations as laborers who were unwilling to capitulate to the drudgery that accompanied the extraction of natural resources. Jobs mining for minerals and ore irresponsibly, polluting streams, felling old-growth trees, and loading extracted products on ships at port docks for international markets are—radical environmentalists would argue later—de facto forms of sabotage upon the essential economies of nature.

Ta-ra-ra-boom-de-ay!
It made a noise that way,
And part of that machine
Hit Reuben on the bean.
He cried, "Oh me, oh my;
I nearly lost my eye."
My partner said, "You're right—
It's bedtime now, good night."

The threat of potential violence escalates, as the second worker takes the place of the first and continues the pattern of sabotage intended to bring about fair working conditions.

But still that rube was pretty wise, these things did open up his eyes.
He said, "There must be something wrong; I think I work my men too long."
He cut the hours and raised the pay, gave ham and eggs for every day,
Now gets his men from union hall, and has no "accidents" at all.

The pat, didactic conclusion to this song drives home the point for listening laborers: sacrificing one's daily wages for the goal of the larger good can bring necessary change.

Ta-ra-ra-boom-de-ay!
That rube is feeling gay;
He learned his lesson quick,
Just through a simple trick.
For fixing rotten jobs
And fixing greedy slobs,
This is the only way,
Ta-ra-ra-boom-de-ay!

In this final chorus, sabotage is named "the only way," the ultimate solution for bosses whose mandates must not be borne. "No compromise," members of the IWW and Earth First! alike would say. The song illustrates that uncompromising attitude well. Joe Hill, who cleverly adapted this song, would himself become an icon, in his life and songs, and today he is remembered as a martyr or sacrificial lamb for the cause of workers' rights.

Joe Hill's image—in painting, poetry, prose, and song—has been a uniting icon for discontented workers in America since he died. In one painting, displayed as a poster in college dormitory rooms and union halls, his face is shaded by whiskers and grief, his head topped by a crush hat (Fig. 4.3). He holds a squeezebox or accordion in one hand, to make the music to accompany his songs, in the other a sheet of paper containing the lyrics of his "Workers of the World, Awaken." That song begins, "If we workers take a notion / We can stop all the speeding trains, / Every ship upon the ocean." He sports a red button on the lapel of his pinstripe jacket and a red background from which he steps.

This image was found on a website entitled "Sabots—toss your shoes into the machine works," a website that itself was an artifact of sabotage, insofar as it stemmed for a while from a site that promotes tourism on the seaboard of North Carolina. Someone hacked the tourism website and sabotaged it. The "Sabots" site has photos of wooden shoes as well as quotations from Elizabeth Gurley Flynn, who wrote a 1915 pamphlet entitled *Sabotage*. Joe Hill celebrated her in the song "Rebel Girl." As if Hill were still alive to oversee the subterfuge of exploitative capitalist enterprises, his image continues to duplicate itself in the emergent media of the Internet today. In a more contemporary painting, originally a 5′ x 7′ banner, an angelic Joe Hill appears in the foreground; above him floats a

Figure 4.3. Carlos Cortez, *Joe Hill*, Linocut, 1979, reprint courtesy of Industrial Workers of the World. ©Industrial Workers of the World Union

steaming pie that functions as a halo and beatifies him. That "pie in the sky" alludes to his song "The Preacher and the Slave," whose chorus belittles Christian promises of an afterlife as so many slices of pie in the sky. In the middle ground, two blockheads or squareheads look on, one of them draped in an American flag, the other in garb of the Christian clergy. The term *squarehead* refers to stupid persons. Much as Karl Marx had dubbed religion an "opiate of the masses," so the Wobblies ridiculed heaven as a kind of treacle, an insubstantial dessert whose assurances worked to pacify laborers to settle for less than they need or deserve. Such pie satisfies the belly for a spell but not the soul.

Even had Joe Hill not died as a martyr to the cause of class struggle, he might be lionized as a powerful songwriter today. Contemporary folk musicians still perform his music and find targets for their hammers in environmental causes around the world, including the late Pete Seeger who founded the Hudson River Sloop Clearwater group in 1966 with his wife to protect the Hudson River watershed. DiPetto pointed out just how times have changed since Joe Hill died: "Today Salt Lake City is home to the Joe Hill House of Hospitality. The State of Utah now says that under their present law, Joe Hill would never have been executed on the evidence presented at his trial" (152). Whether Utah officials were sincere in this statement or simply self-exculpating is anyone's guess.

THE DOUKHOBORS OR SONS OF FREEDOM

Taking cues from American revolutionaries a century before, the Russian Doukhobor immigrants to Canada sabotaged via bombings, arson, and acts of public nudity. Their nude bodies massed in town squares, or kneeling before their own possessions they had set afire, made them iconic international saboteurs of social composure. Their naked protests constituted a kind of psychological guerilla warfare. Escalating the warfare, they later began to target Provincial government buildings to inflict heavier economic damage.

Their objections to mother Russia's government had been many. They denounced military conscription, priests and rituals, the divinity of Jesus, indulgence in alcohol or tobacco, secular government, and private property over collective ownership of land. The culminating gesture of their outrage took the form of a risky public demonstration: piling rifles in a public common, they burnt them in 1895 to protest the military draft of Czar Nicholas II, thus embodying their objection both to enrollment for military service and to compulsory contribution of money for war. In their resultant persecution and diaspora, they suffered starvations and death by exposure. Prince Peter Kropotkin (1842–1921), the Russian aristocrat, zoologist, evolutionist, and anarchist, suggested emigration to Canada as a solution for the Doukhobors—the same Kropotkin who would inspire Edward Abbey in his 1950s MA thesis. Support for Doukhobor immigration to North America came also from Count Leo Tolstoy, who dedicated royalties to their exodus in 1899, as did the Society of Friends, thus connecting 1950s Canadian Doukhobors to the 1650s Quakers, who found themselves running afoul of Puritan hegemony and trying to sabotage it by flouting anti-Quaker orders in Boston. For refusing to recant their beliefs, Quakers were hanged (Krawczynski xvii). In philosophical debates, a fine line divides civil disobedience from vandalism and mild kinds of sabotage (Perry 15; Welchman).

Doukhobor reasons for marching nude evolved in their years in the New World. Initially, "the gesture was meant to signify Doukhobor rejection of material possessions" (Adelman). For the Sons of Freedom (the most radical of the Doukhobor sects), the nude marches aimed "to bring believers in closer contact with God." Other commentators have even suggested "they removed their clothing in order to emulate the innocence of Adam and Eve before the fall" (Androsoff 240). After police began to arrest them, though, their motivation changed to "defiance of authority. Thereafter, Doukhobors stripped regularly. Upon the sight of an approaching police patrol, whole groups would undress," an activity that "terrified authorities" and "created a sensation in the Victorian press." The operative word here is "terrified." Theirs was psychological guerilla warfare. Still, *terror* as a term is too loosely used, lightly applied, and coherent definitions of it are in short supply. As "the largest single mass migration in Canadian history" ("Doukhobors"), and as a force destructive of social order, the willful Russian pacifists gave Canadian authorities fits.

With the extreme Sons of Freedom Doukhobors, the gulf between the citizens and government widened and the forms of sabotage grew grimmer. Backlash against forced residential schooling of Doukhobor children in New Denver, BC, from 1953 to 1959 is now seen as the leading cause of their radicalization ("Lost Childhood"; Androsoff 345). Those true believers held that Canada glorified violence and militarized their children; thus they refused the dictates of public education, exposing their children ironically to higher levels of violence at home when their bombs began to go off and their buildings burned. The RCMP named such acts "depredations" (352), which is another locution that compounds the stew of language applied to sabotage and its iconography. Images of mangled bridges, charred courtrooms, and obliterated graves contradict the pacifism that caused Doukhobor ancestors to pile their rifles in a public square and set them ablaze.

Analyses of Doukhobor customs divide between exculpatory accounts (written often by Russian-Canadian partisans themselves) and sensational narratives that place blame and focus chiefly on the violence. A final image relevant for this discussion presents the immigrants in the yokes of their own plows and tilling the fields by the sweat of their own brows. An accepted rationale for doing so: to relieve the animals from the servitude of the yoke. One probably apocryphal reason for relieving the draft animals is that leader Peter Verigin convinced them to free their brethren-animals of cruel labor. Ashleigh Androsoff explains the practice more plausibly: "Each Doukhobor village, home to about a hundred people, only had one team of oxen or horses. These animals were required for transportation, and could not be spared for plowing." The rationale in Androsoff's good analysis wears an economic complexion instead of an ethical one: "Some of the senior women remembered a tradition whereby young Russian women hitched themselves to plows to start the first furrows in a new field as a gesture of respect to the earth, in the hopes that the land would prove fruitful. The women suggested that they could reenact this tradition as a means of getting their gardens started" (83). The image enforces stereotypes of husky Eastern European matrons taking matters into their own hands and making necessity the mother both of invention and tradition. Such "respect to the earth," as it foreshadows the primitivist impulses that would fuel some later environmentalists, may have proven instructive to animal-rights activists that practiced forms of radical sabotage a decade later on behalf of their non-human brethren.

Another apt interpretation of the human plowing, though, connects Doukhobor practices with the guerrilla theater practiced by Earth First! as late as 2013 to protest affronts to the planet including hydro-fracturing (fracking). Photos of Doukhobors posed in reverential sacrifice before their homes they had torched, or posed nude beside piles of smoldering possessions, make it plain that they recognized an image event when they created one, that they appreciated a "photo-op" a quarter-century before the trendy phrase came into play. They had been so heralded, had garnered so much attention from media, that their acts of hitching to the plows might have been communal theater, cultural stunt. Such a suggestion, though, by no means diminishes the efficacy of their effort or the sincerity of their cause. As DeLuca observed, the proof is in the pudding of perception.

EDWARD ABBEY

Born in the state of Pennsylvania in 1927—just before the Great Depression, an era when the language of class struggle was more visible for Americans—Edward Abbey learned early to distrust authority. To complete his master's degree in philosophy at the University of New Mexico, after an honorable discharge from the U.S. Army, he wrote a thesis on eighteenth-century European anarchy. That intellectual interest remained constant in the novels he began to pen in the 1950s. His fictional experiments concerning the strategy and practice of anarchy culminated in the 1975 novel *The Monkey Wrench Gang*, a book that did more than any other books before or after it to bring to the American consciousness a serious contemplation of sabotage and its leading rationales today.

In virtually all of his writing, Abbey expresses his stout opposition to capitalist control, thereby allying himself with the early Luddites. *The Monkey Wrench Gang* and *Hayduke Lives!* paint bulls-eyes on the military-industrial complex that Abbey perceived as devastating the outdoors he loved so much. Industry is the antagonist of his books, industry abetted by government. What concerned him especially were moral imbalances created when technical specialists manage a society. In the essay "A Writer's Credo," from *One Life at a Time, Please*, Abbey disclosed one of his chief purposes as a writer and a private citizen: "To oppose, resist, and sabotage the contemporary drift toward a global technocratic police state, whatever its ideological coloration" (177–78). He thus equated industrialism and totalitarianism. Nor was his antipathy to industrialism confined to rhetoric and artifice. From his earliest journal entries until his last book, entitled *Hayduke Lives!*, he expressed a consistent desire to challenge and act against the powers of industrial control. A physical outlet is essential both to keep the soul from the corruption of inactivity and to give balance to the lassitude of intellectual pursuits. Convictions must be not only actively realized, by destructive force if necessary, he believed. Attacking industrial excess is a middle ground between surrendering to the proponents of mechanization on the one hand and practicing a violent anarchy on the other.

Before Abbey's influential later novels, only historian Lewis Mumford had published sustained critiques of technological processes—"technics," he called them—beginning in the 1930s and culminating with the monumental two-volume *The Myth of the Machine*, the first sentence of whose Prologue expresses misgivings about industry's impact on environment. Developing his theory of ecodefense, Abbey mated hatred for industry and an overweening government with refined knowledge of natural history and anarchy. In his essays he invoked Mumford approvingly and called *The Monkey Wrench Gang* "four little humans against the glittering Tower of the Power Complex, Mega-Machine. Should dedicate it to Lewis Mumford. He'd disown it. Not his type" (*Confessions* 237). Foregoing a dedication to Mumford, then, Abbey dedicated his novel instead to the apocryphal leader of the Luddites, Ned Ludd. The actual quotation that comprises the dedication—from the earliest citation of *Luddite* (1847) in the *Oxford English Dictionary*—might have been too lengthy and dated for Abbey's purposes. In any case, he supplanted it with an adapted or composite quotation from a spurious dictionary. He slyly credits the "Oxford Universal Dictionary" with a definition of Ned Ludd, who was "a lunatic living about 1779, who in a fit of rage smashed up two frames belonging to a Leicestershire 'stockinger'" (5). Edward Abbey rarely could keep a straight face, no matter how grave his topic or whom his writing was aiming to address.

Abbey's distrust of industrialism and technocracy developed early in life. His thrity-eight-year journal seems to have been written for his eyes only, and hence is less apt to indulge in the contrarian aesthetics that make his work a problem for some audiences. Biographer James Bishop Jr. agrees with Wendell Berry that Abbey was an autobiographer, that his life and writing mesh, and that he was most at home when "dismantling the instruments terrorizing nature through sabotage—whether allegorical, metaphorical, rhetorical, or real sabotage" (125). Earth First! co-founder Dave Foreman offers another angle and a more intimate understanding. Like the Mudhead Kachinas of the Zuni Pueblo tribe, among whom Foreman once lived, Abbey relished irreverence: "While the most sacred rituals were being performed, the Mudheads were cutting up, making fun of everyone—as if a nun were mooning the faithful while the Pope gave his Easter blessing." Not only was Abbey accordingly "the Mudhead Kachina of the conservation movement," in Foreman's phrase (174); he was also a Mudhead as artist. In this sense he recalls Joe Hill, the itinerant minstrel and Wobbly rabble-rouser of an earlier generation, whose sensibility was also contrarian. Both made it a habit to shatter the icons of the capitalist status quo.

Distrust of technology and the sensibility of the saboteur are rife in Abbey's essays and novels, but they are most tellingly expressed in the journals. One can almost hear the sound of his teeth grinding between the lines when he wrote to himself in 1955,

> Tolstoy's prophecy has been fulfilled: The advance of scientific technology has aided more the tyrant than the free man. Example: Every man can own and operate a rifle, but tanks and planes and atom bombs are and must be the property of the State (126).

Following Tolstoy, that is, Abbey equated despotic rule with the ability to command technology, a fear that forms the thinking of today's Neo-Luddites, who believe the widespread adoption of technologies facilitates the centralization of power. Almost certainly Abbey would have disagreed with Karl Marx, who foresaw technology as a viable means of liberating labor from the shackles of capital. Marx, we remember, discredited the Luddites of early industrial England for failing at first "to transfer their attacks from the material instruments of production to the form of society which utilizes those instruments" (554). As if responding to Marx, Abbey asserted in his journal in 1956, "Not socialism, not capitalism is the enemy but—industry and technology carried to excess, to and beyond the point of madness" (139). The passage suggests that technocracy can develop, madly and excessively, apart from all political persuasions. Seven years later, twelve years before its actual publication in 1975, he projected a book to be called "*The Wooden Shoe Gang* (or) *The Monkey Wrench Mob* (a novel about the 'Wilderness Avenger' and his desperate band; sabotage and laughter and wild wild fun)" (185). A belief in the efficacy of ecotage or sabotage provided a fertile middle ground for him to explore in his novels. If as a citizen he was helpless before the industrial juggernaut, at least as an artist he could proffer some vicarious satisfaction.

Abbey's run-ins with industrialism were not exclusively vicarious. Most readers would agree that "if there was one transforming event in Abbey's life . . . it was the construction of Glen Canyon Dam" (Bishop 122). Much has been written about this dam that drowned what many people believed to be the world's most beautiful canyon and created the massive Lake Powell. Abbey returned to it again and again in his work. But medical technology rankled him just as much. In a journal entry for 1957 he raved against "the MAD SCIENTIST—once a baroque villain, now a dominant, honored and commonplace figure in modern life" (139). In ways he never shared publicly, this figure gained an upper hand when Abbey watched the father of his second wife, Rita, die slowly in a New Jersey hospital in 1962. "It's like some kind of nightmarish experiment: the mad scientist, once a creature of fiction, now lives everywhere, and dominates our lives" (175). Prosthetics, drugs, tubes, tents, and tools—the applied sciences run amok—robbed Rita's father of dignity and compounded Abbey's recollection of the knee surgery he himself underwent in 1960, a memory to which he devoted twelve published journal pages. When his third wife died of leukemia in 1970, he reenacted it and confirmed his technophobia by decrying "the repulsive degradations of the hospital routine" he witnessed again (225). Violent death in wilderness solitude, he admitted, was preferable to "the ultimate horror: death in a hospital bed, surrounded by the engines of medical technology and technicians making notes" (*Journey* 217). A common element in these garish scenarios is the mute and dehumanized attendants dedicated to effecting technical ends. Biographically speaking, the author's fierce autonomy had to capitulate in his life to a technocratic will to power. In his own death he would not capitulate, however. Instead he died at home, as he intended, and was buried in the desert by friends—stiffing the American way of death, with all of its attendant technologies and costs.

Abbey's journal likewise offers glimpses of the biographical bases for his most germane and famous novel, *The Monkey Wrench Gang*. The influence of this book on the saboteur subculture may be impossible to assess, but one may measure its impact in part by noting the way the title has been commonized—as a verb, *to monkeywrench*, and as a noun, *monkeywrenching*—among activists and in books by Foreman, Manes, and Scarce. A journal entry from 1966 has Abbey resolving "to raze more billboards" (*Confessions* 209), an activity that readers associate most closely with Doc Sarvis of the novel, who is first seen with "empty gas can banging on his insouciant shanks" as he torches those highway eyesores (17). Confessing to razing billboards suggests that Abbey anticipated no scrutiny of his journal, no prying into his life. This confession also strengthens the evidence that he acted out the capers of his characters before he committed them to print. Wendell Berry's assertion that Abbey was an autobiographer, not an environmentalist (Hepworth and McNamee 12), gains support in the posthumous film entitled *Edward Abbey: A Voice in the Wilderness*. There his friends Charles Bowden and Jack Loeffler testify to his "field research" for the book, a point that Loeffler has developed in print and that Abbey's friend William Eastlake also confirmed (Hepworth and McNamee 20). In this case, "field research" is a sly euphemism for actual sabotage. As researchers, they were testing out what most efficaciously might stem the commercial tide devastating the American Southwest, its deserts and its canyons, its undammed rivers and clear skies.

Finally, a journal entry tells readers how Edward Abbey anguished over matters of sabotage: "Thermite, dynamite, sugar in the gas tank. Falling billboards, falling in flames. D.A.C.—direct action committee. What else can we do?" (188). Out of context, the passage has a lyric rhythm and tone and could serve as credo for American sabotage, but in context it is clouded by the darkest despair. He wrote it in 1964 in Hoboken, New Jersey. There he had gone to be with Rita to try to patch up their ailing marriage. He was homesick for the desert Southwest, doubtful that the domesticity would persist, drawn by "the sweet unbelievably lovely girls" around him, troubled by a chronic lack of money. Outside of a year in Scotland on a Fulbright fellowship in 1951, this is the only time after 1947 that he would live away from the Southwest. A thrall to forces beyond his control—a bad marriage, love of a distant land, a raging libido, and economic straits—he was thinking in icons. The iconography of the passage suggests he wanted to sabotage those frustrating circumstances of his life to bring his despair to a decisive if irrational end.

Eight months after writing this telling passage, he would flee New Jersey and return to the Southwest, to Utah, confessing in his journal, "Bolted. Left wife and kids and job for exile in the desert" (194). Similarly displacing his anxieties, Henry Lightcap opens the autobiographical novel *The Fool's Progress* by taking his .357 Magnum pistol and shooting his refrigerator, domestic emblem of his failed relationship with partner Elaine.

EARTH FIRST!

Abbey was an influence on many environmentalists, especially Dave Foreman and his fellow members of Earth First! Motivated by Abbey, those activists set about monkey-wrenching the tools of industrial forestry and extractive industries of many kinds. (In the literature of environmental activism, to extract is to wrench from nature for profit without planting, to extort from public waters and lands. Threatened salmon are one example, as are waters polluted by factories, timber taken for private gain, and grasslands grazed to dust by ranchers' herds.) The monkey wrench became a key symbol in the iconography of sabotage. The literal monkey wrench is a spanner with a movable jaw that is often toothed to grip more securely. Jones notes the wrench is "the weapon of choice for Charlie Chaplin's Little Tramp in *Modern Times*" (198), but viewers of that movie will remember that the wrench also served to enslave him on the assembly line. As a symbol of wreckage and resistance, it represents the force of will that prompts some activists to risk their lives by disabling machines with tools. For a century the word has been used as a verb meaning to disrupt, obstruct or spoil (*OED*). Yet another usage of the noun has evolved: *monkeywrenching* (see Abbey's "Forward!" below). In the same way that the old Luddites spoke of throwing a sabot into a machine, so their followers in America spoke of throwing a wrench in the works. Just as Joe Hill's pitchfork served the frustrated farm worker in his song, so the monkey wrench came to serve offended environmentalists who saw the public forests and deserts being plundered. As a form of tribute, the Sea Shepherd Conservation Society named one of its vessels *Edward Abbey* (now *Sirenian*), a boat that is featured in the reality TV series *Whale Wars*. The entire fleet of Sea Shepherd vessels is named Neptune's Navy, an identity that hearkens back to that god of Greek mythology whose wrath rose up to defend the sea and its creatures. The ships advance to ram, obstruct, or otherwise sabotage whaling and driftnet fishing craft that operate in the international high seas illegally. The Sea Shepherds and founder Paul Watson have seen their documentary TV series running for six seasons on the Animal Planet cable channel.

Earth First! borrowed from its radical predecessors. The environmental activists co-opted the black cat (sabo-tabby) of the Wobblies and stuck icons of it on logging machinery as a warning to those who owned and drove the machines. For operators in U.S. forests, it must have been unnerving in the 1980s (when Earth First! was most active, most militant, and public-lands logging was at its height) to climb aboard their bulldozers and discover these adhesive icons on their windows or instrument panels. The desired effect was to get them talking and contemplating the consequences of their actions. Mike Roselle, one of the founders of Earth First!, borrowed from anti-war and Black Panther icons to create his green fist logo (Fig 4.4). This icon or emblem shows a hand from the wrist up clenched in a fist, in determination, and painted in the color green to represent the environment. That green fist touts a power to the planet rather than a power to the people, as the upraised Black Panther fist had done. (The Black Panthers were militant African-American activists who flourished from 1966 to 1982.) Vigilant cats and upraised fists, whether black or green, conveyed the necessity of property destruction to achieve political ends. The Wobblies also taught Earth First! the tactic of tree spiking (which is now giving way to "tree pinning" to prevent both visible detection and audible detection when pounding in spikes). This form of sabotage involved guerillas or undercover activists going into forests

Figure 4.4. Earth First! Fist, reprint courtesy of Earth First! All images from EF! are in the public domain except those from R. Crumb.

and driving spikes in trees that were targeted for "harvest." The spikes made of tough substances, usually steel or porcelain, would break a saw blade in the mills where the trees were trucked after being cut. To save unwitting workers from injury—and of course to save the trees themselves from harvest—activists would transmit messages to authorities that a given target stand of trees had been spiked. The Wobblies used tree spiking as a device to strike out against bosses of logging outfits and mill owners who refused to address their labor demands. The demands of Earth First! were simpler to understand and comply with: stop cutting threatened and old-growth trees. Their demands were unselfish, other-directed, and entailed no profit motive.

The word *ecotage* was coined near the first Earth Day in 1970 and used widely in the *Earth First! Journal* and other publications. In recent years it entered the *Oxford English Dictionary*: "Sabotage with an ecological motive; terrorism carried out against businesses and institutions which are considered to be causing unnecessary damage to the environment." Note that "terrorism" in this definition significantly does not target people, only corporations and institutions. The "ecological motive" above alludes to vindictive respect for that interconnected web of human and nonhuman lives that rely on nature. The word *ecotage* now yields hundreds of hits on the World Wide Web. Callicott and Frodeman handle the concept of ecotage in more ethical terms: "It is often deployed by radical environmentalists who consider all species to be intrinsically valuable (and/or the earth to be sacred in some way) and believe that humans have precipitated an environmental crisis so grave that civil disobedience and even sabotage are morally permissible—if not obligatory—tactics in arresting the destruction of nature" (287–88).

The moral dimension of this definition takes the concept to a different philosophical level by including the word *obligatory*. For some planetary caregivers, Callicott and Frodeman suggest, for those afflicted (or perhaps gifted) with compassion, there is a compulsory and heartfelt mandate to act on Earth's behalf, on the behalf of other species, even if that means acting out, acting extra-legally, and sacrificing one's own freedom and welfare. In the same way that artists can't help but create, so ecoteurs can't not strike out to cripple a system that they see as innately corrupt and stacked against all beings but human beings.

Like the Wobblies, Earth First! did its work most potently in the American West, on farms, mines, logging, and construction camps—industrial operations or trades that were hardest on the environment. Few cases of ecotage ever got publicized. Activists were interested in dodging the law. So was capital, "for resource firms are anxious not to give their insurance companies another reason to raise rates in an industry already beset by safety problems" (Manes 10). Acts of ecotage usually were covert and surreptitious.

In their rhetoric, too, the early Wobblies and followers of Earth First! discovered mutual ground. Bill Heywood's 1915 speech in Chicago, which brought into being the IWW, integrated the phrase "no compromise" (Dubofsky 46). Earth First! modified that phrase to say, "No Compromise in Defense of Mother Earth." The physical alliance of these historically disparate groups began with Redwood Summer, a 1990 event that called successfully for common cause between those who valued the timber industry and those who wanted to rescue the few remaining old-growth redwoods. That collaboration was something of a historical coup; environmentalists and workers had been at odds for many decades. Some simpleminded environmentalists claimed that workers got drunk and raped the land, while industrial interests told workers that greens locked up forests and robbed them of jobs. Under such rhetorical conditioning, resentment grew.

Earth First!er Roger Featherstone tried to heal the rift. Writing in the IWW newsletter with rhetoric meant to mobilize, he noted, "A blow to save wilderness is a blow against the assholes that are screwing the workers" (1). For more than a century, corporations had enacted boom-and-bust cycles in local economies. And so Featherstone asked readers "to challenge the mentality of slash-and-burn and rape-and-run." Eco-defender Featherstone called himself "kind of a union organizer for grizzly bears" (2). Most workers can't afford alpine condominiums or ocean-side vacation homes. They appreciate how public land is "a working person's dream," Featherstone wrote (2). To bring this dream to light, both sets of interests, the blue collar and the green, saw a need to combine their forces and take to task the formidable "guardians of the status quo—cops, courts, preachers, politicians, and the prostituted press . . . ." That alliance continues to this day, even if now offset by government clampdowns and by a diminished harvest (so-called) of trees on national forests, due to more-affordable forest-product imports.

Both groups—the Industrial Workers of the World and Earth First!—had iconic martyrs. Joe Hill, hanged, was immortalized in visual imagery and song. Wobbly Wesley Everest, beset by patriotic Legionnaires, was chased, cornered, castrated, and lynched in Centralia, Washington, in 1919—an image that served Dos Passos in his novel *1919* and might have served as model for the demise of Joe Christmas in William Faulkner's *Light in August*. Earth First! got its martyr in Judi Bari. She and Darryl Cherney spearheaded Redwood Summer, the 1990 action that threatened timber capital in northern California by mobilizing workers and environmental protestors. One day, "Judi Bari's pelvis was blown apart by a pipe bomb in Oakland" (Zakin 10), and Cherney suffered lacerations and a shattered eardrum. Bari later died of breast cancer, but Cherney still is strumming, singing, and organizing. His allegiances to Earth First! grew to include the cause of the striking Kaiser Aluminum mill workers in Spokane, Washington, a city with a long Wobbly history. There in 1998 he came to sing songs and to found a coalition that briefly built a productive common cause. Earth First! partisans claimed that industry goons had planted the bomb that nearly killed the two activists, while industrial interests alleged that Bari and Cherney were carrying the bomb. Bari and Cherney sued the FBI and local police for false arrest and illegal search and won a $4.4 million damage award.

One human icon reviled by both EF! and the IWW is the corporate raider Charles Hurwitz. Chairman of the board, chief executive officer, and president of MAXXAM Inc., the Texas-based millionaire had saddled the public with $1.6 billion in the 1980s savings and loan scandal in America. Later Hurwitz bought Kaiser Aluminum, idled its workers, tried to plunder their pension fund, and then sold government-contract energy back to the people at a tremendous profit. Hurwitz's Pacific Lumber Company was under protest by Bari and Cherney when the car bomb went off. T-shirts, websites, posters, and ads in magazines—all emblematized the face of Hurwitz, who had tripled logging rates on the world's largest private stand of redwood trees in the United States. Many of the foes of Hurwitz, disparate sympathizers with radical environmentalists and with the idled aluminum laborers alike, succeeded in sabotaging the 1999 meeting of the World Trade Organization that took place in Seattle, Washington. That protest—involving smashed windows, disrupted meetings, graffiti, and more—laid the groundwork for international alliances of protestors who have worked also to sabotage the dominance of corporations in those trade pacts that regulate developing countries and developed countries alike.

Ecowarriors, like their predecessors in England and in America's earlier decades, borrow from Wobblies "the iconography of the advancing proletariat" (Buhle and Sullivan 135), but they innovate by putting that iconography to work for conservationist ends. In this variation on a sabotage theme, the interests of ecosystems and other species now take precedent over self-interest. Where the Wobblies "hurtled themselves fecklessly into settings where they seemed certain to end up martyrs" (135), agents of Earth First! and the Earth Liberation Front did night work that allowed them to perform sabotage with less chance of getting apprehended. Part of a connected ideology, these seemingly disparate interests groups have left "a collective visual legacy," "a political calligraphy" that "has made its way into the American psyche" (Buhle and Sullivan xiii, xvi).

HUNT SABS

Arguably the most risky form of sabotage is that practiced by "hunt sabs." Activists who sabotage U.S. hunts put themselves in the line of fire. They come between the guns and the game. Before foxhunting was banned in Scotland, England, and Wales in the first decade of the millennium, hunt sabs deployed leaderless resistance and direct action to confuse or delay horses and hounds. Their tactics included wiring gates shut to impede the horsemen, blowing horns or shouting loudly to confuse the hounds, spraying or rubbing false scents like garlic water, and more (Vale 16). John Prestige founded Hunt Saboteurs Association (HSA) in 1964 in Brixton, England, after being assigned in 1963 to report as a journalist on a Devon Somerset Staghound hunt, where he witnessed a pregnant doe get driven into a village, cornered, and killed (Stallwood 82). Supporters of HSA more recently have taken up joint operations with the Sea Shepherds in high-seas activism, impeded the culling of badgers and seals, and obstructed hare hunts. "The initial Sea Shepherd campaigns included ramming whaling ships, intervening to prevent the clubbing of baby seals, and the cutting of drift nets to prevent dolphin deaths, actions that were later labeled by the Federal Bureau of Investigation in the United States the first cases of ecoterrorism" (Callicott and Frodeman 287–88). In addition to the exposure to danger posed by the presence of guns and clubs, hunt sabs have suffered broken bones and beatings. In 1973 the Band of Mercy, later renamed the Animal Liberation Front, began perhaps the most daring and capable campaign of ongoing sabotage in history when it turned to laboratory arson to stop vivisectionists.

One of the enduring credos of the environmental movement and of ecocriticism, "Think globally; act locally" urges planet caregivers to respect the welfare of the whole globe but take action in their own bioregions, cities, and towns. Abiding by that dictate, I take up as a case study the sabotage of wolf hunts in the Northern Rockies where I reside, and in particular a document produced by Earth First! titled the *Wolf Hunt Sabotage Manual*.

Extinct in the region since the 1930s, wolves were reintroduced to Yellowstone National Park in 1995. As the population grew, they spread out of the park and started to kill cattle and game animals. Ranchers and hunters lobbied legislators and succeeded in implementing hunting and trapping seasons for wolves in neighboring states. Those hunts have triggered battles in the culture wars over natural resources. Scientists and citizens disagree whether wolves have restored an ecological balance to Yellowstone that was lost with the extermination of top-level predators in the last century. Scientists and citizens also split on whether wolf populations have reached a critical mass adequate to withstand "harvests" without disrupting ecological balances and setting back progress made in the federal program to restore wolf populations as intended. Enter Earth First! to the fray.

The wolf has been an iconic choice for Earth First! since the group's inception in 1981. Dave Foreman and other of its founders and followers have taken the wolf howl as a signature and posed for dozens of photos with throats open and heads thrown back. One of the foremost among EF! slogans has been "Howling for Wilderness," a slogan that cleverly inverts paranoid Puritan verbiage about the "howling wilderness"—the great outdoors that their mythology peopled by heathen Indians leagued with wild beasts and Satan himself. The return of the wolf, whether reintroduced or self-repopulating in its former range, was too good to last. Celebrations were rash. Wolf trapping and hunting returned to Idaho, Wyoming, and Montana. Hunters keen to kill wolves organized rallies and posed for photos.

In reprisal, hunt sab philosophy from Britain a half-century earlier came into play in the Northern Rockies. What changed were the tactics. Instead of wiring gates and blowing horns, hunt sabs in the remote and wild northern Rockies needed to behave more like guerillas and anticipate greater prospects of danger to their property and persons. Wolf hunters and trappers, unlike foxhunters, invariably carry guns. In the culture war over wolves, too, they are out for revenge: against government agents they identify as having overreached by reintroducing wolves, and against environmentalists whom they perceive as encroaching not only on their recreation but also their livelihoods, for trappers sell the pelts of wolves and other furbearers, and outfitters guide well-heeled hunters for a living. To add to the mix, some hunters simply view the deer and elk as theirs by entitlement and despise the wolves as competition for venison on the hoof.

The writers of the Earth First! *Wolf Hunt Sabotage Manual* adopted the pseudonym of "the Redneck Wolf Lovin' Brigade," a time-honored habit of anonymity by contributors to the *EF! Journal*. For the practical purpose of concealing their identities, the writers of the eleven-page, illustrated manual claim a kinship with their opponents: "We are hunters and proud of it" (1), thus underscoring a major difference from the saboteurs who succeeded in repealing fox hunts in Britain. Here in the Wild West, one may stake a rhetorical claim as a redneck progressive and gain greater moral authority than as a non-hunter or vegan.

The iconography on the opening page of the *Wolf Hunt Sabotage Manual* is a pair of saboteur's tools: a monkey wrench and a stone hammer, both of them crossed beneath a photo of a howling wolf (Fig. 4.5). The hammer is particularly symbolic. Unlike Enoch's hammer, which was a forged tool made by a modern craftsperson, the EF! hammer lashes together two poles for a handle and a grooved stone head. Such crafting, used by paleoamericans, is called hafting. Sinews lashed the handle to the stone. While hafting represented a great advancement in the development of technology for its time, today such a hammer is a decidedly primitive tool, raising the question as to why EF! would adopt such an icon. The best answer is that it intimates anarcho-primitivism as an icon. That is, it represents anarchy in its destructive force and defiance of authority, and it represents primitivism in its traditional construction—its rudimentary sinews, stone, and wood. The monkey wrench, with which it forms the iconic cross, has represented sabotage for a century. Anarchy, sabotage, technology, and primordial force combine in this one composite icon.

Anarcho-primitivists critique agriculture, civilization, and domestication. They hold that such forces estrange human nature and make a feral state desirable by contrast. They endeavor to restore wildness both in human behavior and in the natural world, as embodied by wolves. They also advocate an anti-technological alignment. Many abide by the saying of Thoreau: "In wildness is the preservation of the world" (often misquoted to say *wilderness* instead of *wildness*). A common credo of both animal liberation and earth liberation movements is "Live wild or die," indicating that a primal wildness is so very vital that to live without it will prove fatal. One way to regain wildness and human health is to engage in combat for righteous causes. An online English comics collec-

Figure 4.5. Earth First! tools, reprint courtesy of Earth First! All images from EF! are in the public domain except those from R. Crumb.

tion entitled *Sabotage,* a project of Vale and Valleys Hunt Saboteurs, set out to depict "the fine art of hunt sabotage, featuring articles on tactics used to disrupt hunts, the police and the law."

The process of restoring the wild in the natural world is called "rewilding," a term coined by EF! cofounder Dave Foreman who also founded and now leads the Rewilding Institute. Rewilding gained credibility in conservation circles after biologists Michael Soulé and Reed Noss defined it as a legitimate conservation methodology, one that would be based on the "restoration and protection of big wilderness and wide-ranging, large animals—particularly carnivores" (2). Their article pertinently appeared in the now-defunct environmental journal *Wild Earth* (1991–2004). All these prongs of wildness and science are intimated by the coupling of the wolf photo, wrench, and stone hammer in the Earth First! *Wolf Hunt Sabotage Manual.* Its subtle evolution or transference—from icon, to philosophy, to science, and finally to conservation praxis and sabotage—is an instance of image politics in ecological relations. Scientists Michael Soulé and Reed Noss might not welcome being affiliated with anarchists and saboteurs, juxtaposed alongside stone hammers and monkey wrenches, nor perhaps would they be grateful for being perceived as misanthropic, a frequent allegation of anti-environmental rhetoricians and activists, but one could say with confidence that they would support most measures to defend wolves.

Delicath and DeLuca coined the term "image event" to name "staged acts of protest intended for media dissemination" (17). Elaborating on the phrase and definition, I would push it further, extend it, apply it to static icons of environmentalism as well. Indeed, I would argue that iconography has distinct advantages over image events as defined above. Although the image event is dynamic, it is also temporal, place-bound, and ephemeral. It is news that grows dated soon after it becomes televised. Ezra Pound famously defined poetry as "news that stays news," and in key respects the same applies to icons of the environmental movement, irrespective of whether they represent sabotage. Although the icon is static, and might seem to suffer those disadvantages, it is atemporal, free of the limitations and the ravages of time; it is placeless and thus transferable across a broad spectrum of activities and applications; above all, it is adaptable and sustainable. It wears well. Its malleability to different interest groups benefits it. Its imagery lends itself to analysis and redefinition as new groups embrace it and adopt it as their own.

As subcategories of visual arguments, image events represent possibility and actuality alike. The stone-and-sinew hammer is an artifact that human ancestors made for survival. As a paleontological actuality, it resides in museums and collections. It exists also as possibility, inasmuch as saboteurs today may wield such crude tools to impede injury to the natural world, to dissuade corporate capital and governments from pursuing detrimental projects, or to thwart the advance of technocracy and ruinous industries in general. Whether made of steel or stone, a hammer swung against an exposed part of a machine will have the same effect. Beyond the literal and the possible, of course, the image gains force from its longevity as a tool used by the human race. As image event the hammer—repeated in dozens of contexts in the three decades since EF! came into being—is both an actual tool that activists manipulate and a metonymy for resistance. As a metonym, it represents the force of overt and covert activism for the wild gray wolf by those who oppose its hunting and trapping. Conjoined or compounded by the monkey wrench—that tool fabricated by machines from factory steel—the atavistic force of the stone hammer gains the delicacy and sophistication of modern technology and industry. A judo artist may appropriate his or her opponent's strength and turn it to advantage in much the same way. As an image event, the compound iconography that complements the *Wolf Hunt Sabo-*

tage Manual partakes of the efficacy of those "staged acts of protest designed for media dissemination" (Delicath and DeLuca 315). Viewers may register the show of force in the image event of the hammer and wrench on a primal level, even if they do not read the entire manual or ever endeavor to thwart a wolf hunt. The compound icon, like the image events analyzed by Delicath and DeLuca, is "an effective tool for addressing the problem of the 'distracted and disinterested' audience" (326). Equal parts defense and offense, the crossed tools dramatize the earnestness of the cause and the conviction of those who use those tools, whether emblematically or actually.

The *Wolf Hunt Sabotage Manual* is richly illustrated and stirring in its accounts of dodging hunters and releasing wolves from traps. Like the book *Ecodefense* that they reference, the authors have honed techniques through field research and experimentation on the ground. They have learned that images are as powerful as words. Their advice on investigating "the fur business infrastructure" (6) is shrewd. Their technical guidance on building a "Deluxe DIY Pole" for freeing trapped wolves is engineering-level in its complexity and sophistication (8). Rhetorically, the manual keeps its eyes on its audience by dropping verbal bombs like "slob hunters," "goons," bloodbath" (1), "assholes," and "grinning yuppie" (3). Such language ramps up the rage and indignation, making plain that the writers are targeting oppressed insurgents or "subaltern counterpublics," in the term of Delicath and DeLuca (324). The juxtapositions of erudition and primitiveness might jar some academic audiences, especially those of us excessively "constrained by narrow notions of rationality and civility," but the manual is a definition by example of the emerging anarcho-primitivism that ethically underscores the subtle iconography of contemporary environmental sabotage. The final words in the manual are "The Wild Beckons and Thanks You," personifying the wolf, restoring it as part of its ecosystem.

DAMNATION (2014)

The first face to be displayed in this feature-length film is a graffiti artist. His name is Mikal Jakubal. Busy in the late 1980s as a radical activist in Bellingham, Washington, he lives now in California and works as an EMT, gardener, and documentary filmmaker. Back in the day, though, he doctored billboards, crafted flyers, and fomented ecotage and mayhem. He created a memorable iconography. His specialty was defacing dams. He added culturally to the toppling of two federal dams that came down between 2011 and 2013.

DamNation is "dedicated to those who work passionately and tirelessly to protect our rivers." Funding came from Patagonia, whose founder Yvon Chouinard appears in the film. A lifelong fly fisherman, Chouinard has a passion for salmon, which have been reduced in the Columbia River drainage to 8 percent of their former glory. Bruce Babbitt, U.S. Secretary of the Interior under Clinton from 1993 to 2001, also interviewed in the film, says some 75,000 dams have been erected in the United States. "That's more than one every day since Thomas Jefferson was in office." Jakubal defended rivers by vandalizing bad dams.

He defended the rivers not for himself, for he is neither a fisherman nor an avid paddler. He defended them to rewild them, to restore wildness to big fish, big water, and himself. He squirms with glee before the camera in telling how he pulled off the one big stunt he is willing to confess. He is willing to confess to that act because the dams had already come down, by federal decree, using federal money. One dam operator with a memory of Jakubal's night spent painting the Glines Canyon Dam crack tells the story.

"We got wind that Earth First! was going to do something that night, so we put an extra ranger on duty." Jakubal got through the added security and committed his caper without detection. That caper, a painted crack with the words "Elwha Be Free," gained one dam operator's admiration. "It was a beautiful crack. The guy was an artist. There was no question of it. And he did that all in one night. It was an amazing feat." Interviewed twenty-seven years later, Jakubal did not want to be remembered for that feat. The operator had another take: "I think that kind of woke up people to the fact that something had to be done."

The film narrator and one of its directors, Ben Knight, says he has a lot to learn from anyone like Jakubal, "who would rappel down a 200-foot dam face in the middle of the night with a paint bucket just to make a statement." Knight asks Jakubal to share "any words of inspiration for aspiring young artists who have a big dam canvas to paint." Jakubal urges activists to grow more ambitious, originate bigger and better image events. While he can't go on record about any monkey wrenching undertaken during the 1980s, I lived in Bellingham during that time and can attest to his radical bravado.

Typifying Jakubal's work is hard. Is he an artist, an activist, a vandal, a saboteur? He had to practice his craft as stealthy as a cat in 1987—first to design and then to paint a huge crack on the face of the Glines Canyon Dam. He had to haul ropes, carabiners, paint rollers, buckets, and sundry other gear by bicycle from his car parked a mile below the dam. He had to perform *Nachtwerk*. If he inflicted only minor economic damage, it sent a major message nonetheless. He chose to combine pictures and words in ways that would mobilize. "'I didn't bring those dams down,' he said. 'We were just one little bitty pinprick in the side of the system, trying to make this idea go forward'" (qtd. in Barcott).

The deconstruction of those dams cost a lot, but a lot less in the long run than letting the largest king salmon species on Washington's Olympic Peninsula die out. "The total cost of Elwha River Restoration is approximately $324.7 million, which includes purchase of the two dams and hydroelectric plants from their previous owner, construction of two water treatment plants and other facilities to protect water users, construction of flood protection facilities, a fish hatchery and a greenhouse to propagate native plants for revegetation" (Olympic). The expensive fish hatchery is an ongoing controversy. It is in the hands of the local Native Americans, the Lower Elwha Klallam Tribe, but the wild and hatchery fish will crossbreed and dilute the genes of the wild strain. The devastation of rivers by our government, the trashing of tribal livelihoods without their say-so, was the most egregious sabotage in Washington State history. Jakubal's caper pales by contrast. Indian ecology relied on the salmon, the wild fish that the dams reduced by 92 percent of their original numbers. Removal of the Elwha dams was "an answer to our ancestors' prayers," said one of the tribal women interviewed in the film.

The film tells several stories besides Jakubal's. Author David James Duncan shows his passion for saving salmon and reads from his book *My Story as Told by Water*. The late Elmer Crow, a Nez Perce elder, can barely restrain his sadness and frustration over the loss of fish and devastation of the free-flowing Columbia and Snake. Folksinger Katie Lee, age ninety-four, shares film footage and photos of the Colorado River before Glen Canyon Dam went in, including photos of her romping nude there. Washington legislator Doc Hastings and Northwest Power and Conservation Council member Jim Yost represent the pro-dam side, both of them explosive with anger that salmon might take priority over human needs. Some film reviewers complained that coverage of the issues is not fair and balanced. But few people are willing any longer to defend such boondoggle dams.

When the two filmmakers try to pass through a set of locks in a lower Snake River dam by kayak, as they had read online they were legally allowed to do, two police and a port security officer harass them and make them reverse course. The confrontation makes for some exciting cinema and some cultural dissonance. Even though the paddlers were approaching public facilities constructed by public tax dollars, the officials accused them multiple times of activities that the government associates with terrorism. The "t" word punctures the air and makes the confrontation stressful and irrational. If tourists approaching by motorboat are allowed to pass, the double standard begs a central question whether terrorists would be more apt to transport bombs by muscle power or motor.

How do we define sabotage in the context of salmon, rivers, Indians, paddlers, and dams? Decommissioning bad dams that have been in place for a century or two is a legal act. It is also a form of destruction that radical environmentalists never took upon themselves. Had they blown up a dam, as the fictional gang in the 1975 Edward Abbey novel *The Monkey Wrench Gang* devoutly wished to do, the environmental movement would not have recovered from the setback delivered by that explosion.

The destruction of bad dams, defunct or destructive dams, is coming about now because their destruction makes good economic sense. Legislators, taxpayers, and engineers come together to undo ancestral goofs. As University of Washington geology professor David Montgomery observes in the film, fifty years ago a conversation about removing dams would have been "unthinkable. Conversations change." Perhaps in decades to come the conversations will change about ecological saboteurs, who may be seen no longer as terrorists, lunatics, and fanatics but as progressives in line with activist-abolitionists one hundred and fifty years ago and gay rights advocates today.

One of the final images, a still, shows Katie Lee holding a bundle of dynamite in one hand, a monkey wrench in the other, her arms crossed. She is not an actor but an activist and folk singer, one of the few living people to raft and explore Glen Canyon before the waters rose. Her tools replicate the crossed hammer and monkey wrench of Earth First! origin, but the dynamite held by someone her age adds an unexpected edge.

CONCLUSION

One of the distinctions discussed above, between the economic and ethical complexions on the face of subversive behaviors, is unavoidable in a full consideration of sabotage. Workplace saboteurs often argued that their own rage had an ethical basis, that laborers' exploitation by grasping capitalists is evil enough to warrant harsh reprisals. There is no fury like righteous self-interest. In the matter of workers' rights, an economic element comes always into play. But one man's exploited circumstance is another's opportunity.

From an ecocritical perspective, it is hard to withhold every shred of respect for subversives bold enough to throw a handful of sand into the proverbial machinery. They have done so not because their personal lots will improve, not because they will be able to eat better or ease into retirement sooner, but because their consciences compel them to acts of compassion for other species and for the environment, often in stark disregard for their personal freedom and welfare, their self-interest and economic stake in the matter.

For more than two decades, the U.S. government has been looking closely at saboteurs who act on behalf of the environment, and the word *ecoterrorist* has been worming its way more and more deeply into mainstream dialogues. In 2010, a so-called ecoterrorist measure (Senate Bill 6566) saw introduction into the Washington State Legislature. In 2012, the federal Department of Homeland Security released a report entitled "Countering Eco-Terrorism in the United States: The Case of 'Operation

Backfire'" about an investigation into a Vail, Colorado, ski resort under construction whose arson resulted in $12 million in property damage and lost revenue. One of the arsonists, facing terrorism charges and life in prison, committed suicide and in a note identified himself as "the most recent casualty" in the war upon the planet by brutal human cultures (qtd. in Callicott and Frodeman 290). Earth First! co-creator Dave Foreman believes the term *ecoterrorist* to be imprecise: "although monkeywrenching might precipitate fear, it was a nonviolent and justifiable tactic given the magnitude of the human assault on nature" (288). A personal anecdote here might offer an apt metric.

People angry over my wife's legal work found our address and vandalized our home. She was working to stop the burning of farm fields on behalf of asthmatic clients who had been disabled by the smoke when stubble burned. We woke one morning before a major courtroom hearing and found our eaves and trees draped with toilet paper. Our car in the driveway was decorated, too, and the toilet paper on the car was soaked in piss. No economic damage befell us in that vandalism or sabotage, an act meant to strike fear and convey the message, *We know where you live and might be back.* What Karen and I felt after that petty event was not terror or anything like it, but anger and a new resolve.

In the legal logic that coined *ecoterrorist,* saboteurs acting in the interest of the environment are being classed as domestic terrorists whose acts situate them legally and philosophically beside those who destroy human lives. The ecoterrorist registry proposed for adoption in Washington State, a registry that was sliding the slope made slick by the Patriot Act of President George W. Bush after September 11, 2001, died in congressional session. But it may be revived, and if passed it would make activists subject to potential registration, like sex offenders are in the United States today: the electronic equivalent of a wrist tattoo. No one ever has been documented as injured or killed in acts of ecotage, and there is no evidence that ecoteurs have ever had such intentions (289). But prevailing political winds in America today, and the rightwing advantage in the American media, where Fox News eclipses all other news programs united in market share, make possible a semantic sleight-of-hand that equates economic saboteurs with Middle East "suiciders." That locution, originated by President George W. Bush, more accurately would be "homicidal terrorists." My beliefs preclude the *terror* element in the imprecise term *ecoterrorism*. In doing so I echo the sentiments, whether radical or mainstream, that the actual sabotage is the terror wrought by government, business, and industry at the expense of species, ecosystems, and indigenous people in the name of progress and economic development.

If iconography denotes "the art or representation by pictures or images," it is well to remember how "iconoclast" is related etymologically. It denotes a breaker or destroyer of images or idols. Joe Hill and Edward Abbey, in their work that enjoys such popular appeal, set about systematically to shatter icons and decry idol worship. The idols they decried were those of capitalism generally and technocracy in particular. By integrating such goals as these in their projects, they originated image events and contributed to a historical iconography of sabotage intended to empower common people. Everyone has a hammer, making moot the conditional mood of Pete Seeger's classic folk song, "If I Had a Hammer." Seeger's lyrics clarify the metaphorical timbre: "It's the hammer of justice. / It's the bell of freedom." The hammer is an icon for the power of oppressed people's conscience to do what they consider right for the causes of social and ecological justice.

FIVE

Rage Against the Machine

Edward Abbey and Neo-Luddite Thought

Edward Abbey's selected journals titled *Confessions of a Barbarian* furnish fodder for readers interested in studying the biographical bases for his fiction, his essays, and his role as a leader of those outlaw activists known as radical environmentalists. *The Monkey Wrench Gang*, most followers believe, generated the goals of Earth First! and articulated them best. The book appeared in 1975, Earth First! began in 1980, and both voice an uncompromising agenda for public lands in the American West. From the book and from its sequel *Hayduke Lives!*, the editors of the *Earth First! Journal* borrowed names and slogans for use on bumper stickers, T-shirts, manuals, and magazines. Edward Abbey spoke at Earth First! rallies, endorsed the group, and wore its clothing in photos. Several books, discussed below, explore the movement and its debt to Abbey, but a more compelling set of connections between Abbey's writings and radical environmentalism remains unexplored: the connections to the philosophies of the still-emerging Neo-Luddites. Adherents to the belief that technology has earned and deserves our distrust and fear, Neo-Luddites and their disparate writings provide a context for a biographical approach to Abbey's fiction and essays.

A hermeneutics of suspicion have dominated Abbey criticism. Every political statement the author made seems subject to repudiation. Even though Abbey praised "the Socialist heritage bequeathed me by my father" (*Confessions* 198), Edward S. Twining argued in 1978 that he is a "radical conservative." Three years later, Jerry A. Herndon indulged in an oxymoron, "moderate extremism," to straighten Abbey's self-conception. Despite Abbey's endorsing Earth First! and authoring the foreword to Dave Foreman's *Ecodefense*, Ann Ronald and Scott Slovic have contended that Abbey is not promoting illegal action. And in the face of evidence that Abbey and his environmental followers have effectively shifted public opinion in favor of the environment, Daniel G. Payne has charged that "Abbey's impact was negligible—perhaps even, as some have suggested, counterproductive" (205). For his part, Abbey took an exceedingly dim view of literary criticism. In 1953, before any of his writings got into print, he opined, "Contemporary literary critics have donned the vestments of high priests and presume to instruct the writer not only in literary method but in what he shall and shall not write about, and in what he shall think about what he writes and does not write" (*Confessions* 109). Rather than regard Abbey with suspicion, I take him at his word, inconsistencies and all, and by studying contemporary Neo-Luddite philosophies below I originate an approach that may offer fresh perspectives on this key element of his fiction and non-fiction. In my path of acquiescence, I would join Wendell Berry, who generously wrote, "'And so his defenders, it seems to me, are obliged to take him seriously, to assume that he generally means what he says, and, instead of apologizing for him, to acknowledge that he is not always right or always fair'" (qtd. in Hepworth and McNamee 14). Abbey at his irascible worst could be nasty, brutish, and short.

The writing for which he is most well-known and admired outside academe is stoutly opposed to technology. The *Monkey Wrench Gang* and *Hayduke Lives!* fixate on the machine culture that he perceived to be devastating the outdoors he loved so much. Technology, not the Mormon character Bishop Love, is the antagonist of these books and many others. The rollicking reprisals against Love—a buffoon whose ruinous machines get wrecked—comically relieve the mood and vary the pace in books whose characters are otherwise grave in opposition to the ruin of their beloved West. For precision's sake, I name technocracy not technology as the chief object of Abbey's distrust. No simple-minded vandal, he was concerned with the moral imbalances created when technical specialists chiefly manage society, which is my working definition of technocracy. Here is a definition that functions by example: demonstrating "how wolf technocracy can run amok," Rick Bass in his book *The Ninemile Wolves* criticizes radio collars when they are attached to endangered species in the wild, for they can taint the experiences of those animals. "Just behind the collar are trappers, helicopters, computers—*knowledge*, instead of mystery" (92). Abbey likewise lamented the loss of mystery in the wild he loved.

Abbey equated technocracy with totalitarianism. In the essay "A Writer's Credo," from *One Life At a Time, Please,* Abbey disclosed one of his chief purposes as a writer and a culture critic: "To oppose, resist, and sabotage the contemporary drift toward a global technocratic police state, whatever its ideological coloration" (177–78). Nor was Abbey's antipathy to technocracy confined to his rhetoric and artifice. He had some basis for judging production. He worked both as a county welfare office caseworker and as a technical writer for General Electric, the latter job the subject of several humorously ironic digressions in essays in *The Journey Home* (99–100) and *Beyond the Wall* (182–83). In the latter book the employer's name gets changed to "Western Electric," probably at his publisher's urging, to avoid defamation charges. From his earliest journals until his final book, he expressed a consistent desire to challenge and act against the powers of technical control. Writing books and essays, though, carping in print, is never enough, for "sentiment without action is the ruin of the soul" (*Beyond the Wall* xvi). Abbey believed that our convictions must not only be actively realized, by the destructive force of sabotage if necessary, but that attacking the tools of technocracy necessitates negotiating the tricky middle ground that lies between surrendering to its proponents on the one hand and practicing a violent anarchy on the other. Abbey's 1959 MA thesis in philosophy at the University of New Mexico, entitled "Anarchism and the Morality of Violence," focused on the anarchists Prudhon, Sorel, Godwin, Bakunin, and Kropotkin. The primacy of Kropotkin in the development of such philosophies comes into fuller focus below.

The high seriousness of Abbey counterparts in American environmental literature—e.g., Rachel Carson, Annie Dillard, Barry Lopez, and Leslie Marmon Silko—offer stark contrast to his work. As Lawrence Buell understated it, Abbey preached against trashing the planet, "but with a raffish panache that unsolemnizes his jeremiads" (300). Abbey, we might say, disrupted his own extended lamentations and complaints with the verbal equivalent of belches, farts, and obscene gestures and jests. Not boorishness but self-effacement might well have prompted his lapses away from polite taste, as corroborated by the many accounts of his shy and quiet personal demeanor. His contradictions and crankiness, his political incorrectness, obscenities, and bad jokes grew out of his steady desire to critique and undercut his own moral thunder, eradicate the taint of sanctification in his work and the gravity of his own environmental alarums—not the gravity of other commentators necessarily. Conventions constantly condition Americans to expect non-stop economic growth, boundless technological expansion, and the degradation of the natural environment. Sabotaging our literary expectations, as Abbey did, accordingly may force us for a moment to rethink the uses we make of the natural world and the expectations we heap upon literary writers and writings, especially the canonical.

Distrust of technology is rife in the essays and novels but perhaps most tellingly expressed in the journals that Abbey wrote. One can almost hear the sound of his teeth grinding between the lines when he wrote to himself in 1955, "Tolstoy's prophesy has been fulfilled: The advance of scientific technology has aided more the tyrant than the free man. Example: Every man can own and operate a rifle, but tanks and planes and atom bombs are and must be the property of the State" (126). Like Kropotkin, the role of Tolstoy in advocating anarchy enters my discussion of sabotage and anarchy below. Following Tolstoy, Abbey equated despotic rule with the ability to command technology, a fear that forms the thinking of Neo-Luddites, who believe the widespread adoption of a technology facilitates the centralization of power. (Compare Jerry Mander: "Computer technology has already produced an unprecedented degree of military centralization" [69].) Abbey swiveled ambivalently between the need to fuel anarchy on some occasions and capitulate to technocracy on others. If as a citizen he felt helpless before the technological juggernaut, as an artist at least he could proffer some vicarious satisfaction.

History has been unkind to the Luddites. Those who wrecked power looms in the English lace and textile industry between 1811 and 1816 are spoken of today as simpletons and reactionaries instead of as skilled laborers bent on protecting their trade, families, livelihoods, and communities. One reason may be the Marxist disposition to defend machinery as a way to liberate workers. "Luddite" itself, adapted from the uprising's apocryphal leader Ned Ludd, is used mostly today as a term of derision or dismissal. And yet these revolutionaries of the English Midlands were effectual enough to necessitate the assembling of some 14,000 troops to quell them. Efforts to resist technology today, even to "unplug" from it, are considered foolish, vain, futile. The Unabomber, Ted Kaczynski, is a murderous Luddite in the media's eyes. Such views ignore the strain of thoughtful skepticism, known generally as Neo-Luddism, that has coalesced since 1990 when psychologist Chellis Glendinning issued her "Notes Toward a Neo-Luddite Manifesto," an article that grew out of her book *When Technology Wounds*. That book is a study of "the increasing threat modern technologies pose to our health and to the survival of the earth" (iv). Computers, televisions, microwaves; pesticides, fertilizers, hormones; nuclear power, internal combustion, electromagnetic radiation—such technical innovations, and the imbalances that attend them, inspire Neo-Luddite thought. Abbey himself eschewed television and computers. He wrote jeremiads against reliance on chemical agriculture and cars. His antipathy to the progress of technocracy is easy to document. The Neo-Luddites who followed in Abbey's wake deserve to be heard out, taken seriously, and assumed to mean what they say, not only for the light that their ideas shed upon his literary works but also for students of American studies. All the elements of ecocriticism are inherent in their writings, too; they need only be sifted and distilled.

A foundational belief, best expressed by Glendinning in her manifesto, is that "all technologies are political" (51). They are never neutral tools. Tacitly agreeing, Abbey invented in *Hayduke Lives!* the outsized and outlandish walking excavator GOLIATH, which is all political. Weighing twenty-seven million pounds, wearing "130-foot shoes" (62), the machine is not built on anything approaching human scale. As parody, however, it is personified. It levels desert terrain, devastates life, doing the will of boosters like Bishop Love who promote the mining industry and tourist trade. This political agenda is most evident in the chapter "Seldom's Nightmare," in which the machine pursues Seldom Seen Smith and plays promotional voice tapes that praise the ways manufacturer "Bucyrus-Erie keeps pace with our world's most urgent requirements" (64). The machine has been consciously structured—like all technologies, in the views of Neo-Luddites—to offer its makers "a surefire method of expanding their markets and controlling social and political thought" (Glendinning 51). In a glaring satire of multinational corporations and their neocolonial patterns of trade that their technology facilitates, the machine mistakenly mixes up its voice tapes and begins to play an in-store ad for a K-Mart "patio dinette set" that has been built, presumably, from the raw materials dragged out of Earth by the machine itself (63). In another book and context Abbey wrote, "such is the nature and structure of the industrial megamachine (in Lewis Mumford's term), which is now attacking the American wilderness" (*One Life* 30). Mumford, of course, was the leading philosopher of technology throughout the course of his long career (1920s–1970s).

These sentiments of Abbey's lead to a second major tenet of Neo-Luddism. In the words of critic Herbert Read in 1955, "'Only a person serving an apprenticeship to nature can be trusted with machines'" (qtd. in Sale 265). While the influence of technology on industrialized people may be an open question, the effects of techno-industrial processes on nature are staggering, as Abbey so well knew. "We live in a time," he wrote, "when technology and technologists seem determined to make the earth unfit to live upon" (*Abbey's Road* 123). In the technocratic worldview, by and large, nature is at best a "resource" and at worst inferior, lifeless, and inert. Not to Abbey, however. In an overlooked dimension of his environmental advocacy, he asserts repeatedly that all of nature is conscious, aware of sense impressions. Aspen trees are "*beings*, alive, sentient, transpiring" (*Journey* 208). He grieves junipers bulldozed for a new roadway in *The Monkey Wrench Gang* and mounts a brief sermon, a gentle jeremiad: "No one knows precisely how sentient is a pinyon pine, for example, or to what degree such woody organisms can feel pain or fear, and in any case the road builders had more important things to worry about, but this much is clearly established as a scientific fact: a living tree, once uprooted, takes many days to wholly die" (76). A related and often overlooked element of Abbey's Neo-Luddism is his support of animal rights. In a passage that should be quoted at length to be appreciated, he pleads for laboratory animals and against their "deliberate torture," which he finds "not merely comparable but analogous," causally connected, to political despotism and tyranny. Like some eco-feminists, he concludes doggedly, "contempt for animal life leads to contempt for human life" (*One Life* 170–71), although feminists who revile him for his retrograde views of women have felt free to disagree. Not only does technology empower us to vanquish nature; it physically removes us from our own industrial consequences. Seated atop a Cat D8, a dragline, or a corporate skyscraper, the technocrat can scarcely witness what he has brought about. On the ground, the impacts are hard to overlook. "I have seen what men and heavy equipment can do to even the most angular and singular of earthly landscapes," Abbey wrote (*Beyond the Wall* 55). He had served his apprenticeship to nature while working as a ranger for the National Park Service and the Forest Service.

A third principle of Neo-Luddite philosophy concerns the effects of corporate technology on third-world people. Borrowing from Mumford, Mander discusses "the corporation as a kind of machine, more persevering than the human actors within it, that operates by its own rules" (287). By the rules of that machine, a human being is little more impediment than an animal, stone, or tree. As Mander illustrates, the management of native villages by unsolicited technocratic powers has thoroughly unraveled many traditional cultures throughout the Americas. Abbey similarly lamented the economic determinism impacting Pueblo Indians whose agrarian economy has served them so well for so long. "When the children and grandchildren of corn farmers, deer hunters, and pastoralists study auto mechanics and computer programming, it seems unlikely that the roots and essence of an earth-based culture can survive" (*One Life* 155). Mander details how introductions of technology-for-profit have prescribed social engineering, harmed dietary intake, altered community relations, implemented cash economies, and despoiled or stolen the farmlands of indigenous people. In his account of a trip to Alaska, among natives who used to clean infants with their tongues and fingers, Abbey wryly noted the sudden proliferation of disposable plastic diapers: "Old ways die easily—they tumble over themselves in a *rush* to die—when confronted by the frills of high-tech civilization" (*Beyond the Wall* 180). Resistance seems futile. Steinbeck's displaced Okies in *The Grapes of Wrath* try in vain to find a way to avenge themselves upon the banking corporation that drives them from their land. Half-crazy Muley Graves in that novel believes that if he can strike the tractor that he sees leveling the shacks, then he can disable the corporation that is repossessing the farmland. Readers feel his frustration.

A fourth element of Neo-Luddite philosophy spells out the need for "resistance." This element, following Kirkpatrick Sale, must be "based on some grasp of moral principles and rooted in some sense of moral revulsion" (269)—which is to say that violence against life cannot be tolerated. Abbey's entire canon and his personal philosophy corroborate this view, for non-violent resistance lies at the center of all he advocated. He made a distinction clearly between sabotage (damage to machines) and violence (harm to living things). Terrorism is defined as an act that puts life in jeopardy, a violation of the Monkey Wrench Gang's first moral: that no one gets hurt. The English Luddites violated this principle when they assassinated manufacturer William Horsfall (Sale 145–47), an act that brought the full force of the law upon their heads. Some Abbey critics allege that the shooting death of the security guard at the conclusion of *Hayduke Lives!* violates the gang's code of ethics, too. But extenuating circumstances moderate that harsh judgment. First, the character that pulled the trigger, Jack Burns, is not a member of the gang and had not learned about or committed to its rules. He is instead a throwback to Abbey's earlier novels and to the Wild West code that freely licensed bloodshed. Second, the stakes had changed greatly between the publication of *The Monkey Wrench Gang* in 1975 and its sequel fifteen years later. Forces of technocracy, devastating the West at an unsustainable pace, had reached fantastic proportions, as exemplified by the GOLIATH dragline. They mandated the tactics of revolutionary violence for Abbey. The gang's enemies at the conclusion of the first novel, remember, were willing to blow Hayduke to bits. By Neo-Luddite lights, however, *Hayduke Lives!* falls outside the pale of acceptability, just as the terrorist acts of the Unabomber do.

A fifth and resounding parallel between Neo-Luddite philosophy and Abbey's work includes, again in Sale's words, a belief that "the edifice of industrial civilization . . . seems certain to crumble of its own accumulated excesses and instabilities" (278). The attentive reader need not look far, either in Abbey's fiction or his nonfiction, to find cataclysmic prognostications. The premise of the novel *Good News*, ironically enough given the title, is that the system has crashed and displaced urbanites are waging war upon rural people and economies. In Sale's elaboration, civilization's "sustained assault on its environment" always has resulted historically "in rebellion, warfare, and dissolution" (278). In the words of Dave Foreman, Abbey's adherent and Earth First! founder, "Modern society is a driverless hot rod without brakes, going ninety miles an hour down a dead-end street with a brick wall at the end" (45). In the words of Abbey himself, "I predict that the military-industrial state will disappear from the surface of the earth within a century" (*One Life* 28). These writers all envision a new society emerging from the collapse of technocracy. For Abbey, this new society will take the pastoral shape of "scattered human populations modest in number that live by fishing, hunting, food-gathering, small-scale ranching and farming" (28). For Foreman, "bioregionalism is what is on the other side of that wall" (45). Sale, who like Abbey and Berry eschews computers to write his books, sees one of the Neo-Luddite tasks today to imagine the shape of "alternative societies" (278) which will resolve, unlike their ancestors, "to fashion their technologies with the restraints and obligations of nature intertwined" (279). The cataclysmic scenarios of the Neo-Luddites share a vision of the future, arising dialectically from technocratic excesses and the ensuing cataclysm, which necessitates learning new ways to love nature and incorporate it meaningfully into society.

Edward Abbey, a Neo-Luddite in thought and deed and artifice, also opposed easy categorizations and most certainly would have resisted such comparisons as these. He was an individual. To pigeonhole him is to take away that individuality. Surely his militancy would alienate some of the growing tribe of Neo-Luddites, which includes Amish and Quakers, anti-war activists, New Age visionaries, and more. His views on guns and immigration and feminist politics would alienate still others among their kindred. Still, it remains worth considering how much Abbey contributed—artistically and polemically—to the growing philosophy of Neo-Luddism. One great contribution of his writing certainly is that he gave tongue to radical and revolutionary notions that otherwise would have remained unarticulated—often thought but rarely expressed in public—like the writing on the wall of the cafe toilet stall that Henry Lightcap visits in the autobiographical novel *The Fool's Progress*, a wall scrawled with the sentiments of the discontented "*vox populi clamantis in deserto*" (136).

SIX

Overtures to Sublimity

Assessing the Bureau of Reclamation Art Collection

The federal government long has commissioned artists who could further its agendas, but rarely on the scale undertaken in 1968. That is when the Bureau of Reclamation began soliciting the aid of painters "to depict the imaginative aspects of the Reclamation Program." The quoted notice, entitled "International Opportunities for Artists" and issued in the art journal *Leonardo*, explains the Bureau's mission: "to conserve and develop the water resources of the arid areas of the U.S.A., the 17 westernmost States, in order to make dry lands arable and to augment water supplies for their expanding population." Today the reasons the Bureau sought artistic aid seem plain to see. A division of the Department of the Interior, which manages public lands, Reclamation was under mounting pressure. Americans had made advances in environmental literacy and were regarding with a jaded gaze the tens of thousands of major dams erected in the West since the agency's founding at the turn of the century. Moreover, after the passage of landmark legislation—including the Wilderness Act of 1964, the Wild and Scenic Rivers Act of 1968, and the National Environmental Policy Act of 1969—the Bureau found itself on the defensive, facing threats of litigation and bad fame.[1] By 1973, when Richard Nixon signed the Endangered Species Act and it was adopted into law, the legislative handwriting appeared clearly on the wall for all to see. The ESA would become the single most powerful legal tool in U.S. history to be wielded on behalf of the salmon, steelhead trout, and other riparian wildlife species threatened with extinction by dams in the West.

If the Bureau's motives for commissioning public art under such historical conditions as these seem self-evident, then hindsight also might explain why journalists and scholars have overlooked the hundreds of artworks the Bureau commissioned—overlooked them even after a scandal erupted in 1990, when many of those paintings were reported to have been lost, stolen or returned to the painters. That the U.S. government was exploiting well-known artists and its large budget, to try to exculpate itself during a bad time, simply must not have seemed newsworthy or adequately significant for scholarly investigation. Recent knowledge and methodologies, though, suggest that the art program—in its peculiar fusion of aesthetics, patronage, rhetoric, and malfeasance—may yield itself to theoretical scrutiny at last and after all. Environmental historian Donald Worster has written, regarding the control of water in the American West, that it is "quite a story, one unsurpassed for greed, deception, violence, foolishness, and unbounded idealism" (346). The "imaginative aspects" of dams, as they came to be enunciated in paint by famous artists and in print by avid bureaucrats, helped the Bureau manage its troubled customer relations.

A brief illustration might suffice. *Flaming Gorge Overlook*—one of forty-six paintings linked in color on the Bureau of Reclamation site—positions a dam project as a tourist attraction (see Fig 6.1). The setting is southern Wyoming. Four of the fifteen featureless people appear to be dressed in prison stripes. All of them face the dam. They ride bicycles, stroll, meditate, practice photography, and paint from easels. The watercolorist, Dong Kingman, a veteran of the Works Progress Administration arts program, was well known by the government.

As in *The Oxbow* (1836) by Thomas Cole (a picture that shows Cole and his easel in the foreground, where he is sheltering from a storm), Kingman may be depicting himself at Flaming Gorge in the act of painting. Given the details that both pieces share—of water, storm clouds, and blasted trees in their left foregrounds—Kingman seems to be echoing Cole, although his landscape is much subdued. Power lines and concrete stand in for Cole's meandering river and vegetation. These imagistic parallels recall a quotation from Aldo Leopold, who wrote in his Foreword to *A Sand County Almanac* that "a shift in values can be achieved by reappraising things unnatural, tame, and confined in terms of things natural, wild, and free" (xix).

ARTWORK DEEMED WORTHY BY THE ESTABLISHMENT

BuRec, a nickname used by employees and detractors alike, commissioned and exhibited paintings for five years, from 1968 through 1973. Some artists responded to ads placed in trade magazines, others heard by word of mouth, and still others received private solicitations from federal agents. The nominal organizer of the project was John DeWitt, whose records of the program were conveyed to the Smithsonian American Art Archives following his death. "From a background in the Provincetown

Figure 6.1. Dong Kingman. *Flaming Gorge Overlook.*Watercolor. Flaming Gorge Dam, Utah. Printed with permission by the Bureau of Reclamation.

art colony and an earlier career as a writer for radio and as a publicist, John DeWitt (1910–1984) unexpectedly emerged in the 1960s and 1970s as the organizer for the Department of the Interior of two immense traveling exhibitions celebrating the American landscape" (Brown 38). His second major project was a bicentennial collection, *America 1976*. Responding to DeWitt's call on behalf of the Bureau of Reclamation were a number of noted artists—Kenneth Callahan, Ralston Crawford, Richard Diebenkorn, Lamar Dodd, and Norman Rockwell. The fee paid to the artists never was made known, though, and DeWitt issued a defense for his use of public funds in the BuRec-published magazine *Reclamation Era*. He wrote in those pages, to substantiate the outlays to his federal overseers, "The commissions offered to the painters were a minimal amount above travel and living expenses given them while visiting the sites of Reclamation projects on sketching trips; the total was far less than the artists regularly receive from sales through their galleries and dealers" (14). The forty artists that signed on to the Reclamation art project benefited additionally from expense-paid tours of dam sites, including helicopter tours. Altogether they produced at least 355 paintings, sketches, etchings, and prints. Using public funds to tour public lands, laboring to discover "imaginative aspects" of engineering feats such as Hoover and Grand Coulee dams, the artists may have been lured by the promise of exposure in the National Gallery show opening and in the traveling exhibition that was to follow, with its seventy-three-page color catalogue. The artists also would have gotten satisfaction from being chosen to produce public art, that is, artworks that are "directed toward official activities: government-sponsored programs."[2]

The resulting 1972 exhibition, like the catalogue that accompanied it, was entitled *The American Artist and Water Reclamation*. Had they not looked much into details, the audiences for the exhibition and the catalogue could be excused for assuming the artists had gravitated on their own toward the water-reclamation projects and the magnificent western landscapes that enshrine them. Nearly half the paintings, in the seventy pieces that toured the nation and came to represent the collection as a whole, disclose no dams or irrigation canals. Awesome scenery takes priority and shapes the more compelling works. Probably to forestall any assumptions that DeWitt was hoping for Soviet-style art—didactic, representational, proletarian—the ad for "International Opportunities" avoided language that might invite social-realist aesthetics only. "It is stressed that nonfigurative or abstract subjects are as welcome as traditional figurative ones." The ad included DeWitt's contact information and probably was written by him. His qualifications to organize the show included earlier work as a publicist. After the show opened in Washington, DC, it commenced an extended tour of the nation. Sponsored in part by the Smithsonian Institution Traveling Exhibitions Service, *The American Artist and Water Reclamation* came in 1974 to a college campus in Spokane, Washington. There, an art critic who reviewed the show for the daily newspaper found very little to admire. "Why state-sponsored art has to be safe, uncontroversial and uninteresting is beyond me," Allegra Berrian mused (16). Ms. Berrian concluded her subdued review with language that, in retrospect, might seem facile and antique, no matter how perceptive: "The best reason to go see the exhibition is to look at some original artwork deemed worthy by the establishment." Government commissions, not private patronage, had generated a purpose-driven art collection. That art focused on the control of nature.

To confirm that public relations motivated BuRec to commission the art, one may look to the article published jointly in 1999 by Leslie Stinger, Bureau intern; and Bobbie Ferguson, "lead technical specialist for cultural resources, Technical Services Center, Bureau of Reclamation in Denver." Written three decades after the inauguration of the collection that would be known as the Bureau of Reclamation Art Collection, the article traced the program's history and speculated about its inception. Whether they openly acknowledged it or not, the article also studied customer-relations management, implying that the art program succeeded as an effort to make over damage inflicted by propaganda that conservationists had spread. At the same time that corporations were learning, agencies were learning as well that they needed to move beyond primitive public relations, beyond the faces of healthy children smiling near churning turbines. As Patricia Paystrup has shown, "Faced with increasing public pressures in recent decades, business and industry turned to a more powerful form of public relations called issues management. Issues management is a strategic planning approach aimed at combating the public policy agendas of activist groups" (179). Stakes were as high at the time for the Interior Department as for the private sector, and public relations still today is considered the runway to customer trust.

On the premise that a picture is worth a thousand words, slick pictorial surfaces of dam sites received higher priority than the emerging scientific research regarding evaporation rates, silt build-up, mud flats, salts, leaching, depletion of aquifers, imperiled species, lost habitats, and the subsidies that the public ends up paying to large farmers. Gerald D. Nash elaborated on the PR crisis facing the BuRec:

> As the need for water in irrigated agriculture declined, only the power head produced by dams was important, and such power could be produced by alternative means. Moreover, economic analyses revealed that dams could waste as much water as they saved. On the Colorado River, which already had a number of dams, additional structures such as Marble and Bridge Canyons threatened to spread the water surface and lead to evaporation of as much as six feet of water annually. Glen Canyon Dam already revealed bank storage absorption of water into porous rock along the waters of Lake Powell, leading to increased silting. (108)

Instead of portraying hydrological debacles, though, the artworks mostly position dams and reservoirs as prime recreation sites, as places to recharge the soul. So confident was the Bureau that its audiences would not see through its mild impostures that it links the article, "Portraits of Reclamation," to its website, along with forty-six of the more glorious portraits inspired by water-impoundment projects. A curator of fine art, the Bureau seems to imply, has nothing to hide.

As Stinger and Ferguson show, though, and as history has demonstrated, the Bureau was not interested in generating lasting masterpieces of art for posterity, nor in making work for starving artists as the New Deal had done, nor in assembling a precious art collection that its curators would watchfully and proudly see through the ages. The motivation was political, not aesthetic; reactive, not proactive. The authors acknowledged that the Bureau was reacting primarily to the conservation movement, whose "members were quite vocal in protesting Reclamation projects that they viewed as detrimental to the natural environment" (48), a claim that suggests the detractors of Reclamation dam projects were relying falsely on perceptions merely. A secondary audience for the article "Portraits of Reclamation," published in a kind of agency trade journal entitled *Cultural Resources Management*, must have been the writers' superiors in the Bureau, whom they sought to satisfy by putting the best face on the art collection.

Between 1968 and 1973, Stinger and Ferguson show, the agency had had to toe a tricky line of audience accommodation. It needed to satisfy a restive environmental community, but it also had to keep an eye on "its traditional clients, the irrigators," as well as those legislators in charge of funding the big dam programs, that is, the "congressmen and by extension, their constituents." Stinger and Ferguson posed a rhetorical question: why did the Bureau choose to use visual art as the device to satisfy its constituencies? Why painting, "instead of, say, community service and educational programs?" The answer seems plain today. Launching social-service programs would have seemed like an alternative to jail time, like reparation for misdeeds. The authors speculated about why the Bureau decided art would fill the bill better than education or service. "Although the thoughts behind the decision are unknown today," they wrote,

> there are possible answers. These include the permanence of art—a work of art continues to communicate long after both the artist and the patron, in this case Reclamation, are gone. There is also the multiplicity of readings inherent in a work of art. Although one message may dominate, there are many other ways to interpret and understand a painting, which allows dialogue to occur. The program organizers may have been aware of the tradition of western landscape painting; commissions such as these would have added another chapter in a recognizable school of painting, adding all sorts of secondary meanings and messages to the artwork. Perhaps the most compelling reason was that government-sponsored art programs seemed to be *en vogue* at the time. (48)

It is difficult to credit the claim that "the thoughts behind the decisions are unknown today." Commissioner Floyd Dominy, who led the agency from 1959 to 1969, a year after the art program started, was alive in 2007 and still granting public interviews as late as 2000 (Marston). A second solid source for puzzling out "the thoughts behind the decisions" would have been John DeWitt's records of the art program, which had been bequeathed first to the archives of the *American Art Journal* in 1987. In sum, the disclaimer leading to a need for speculation in the article seems insincere. Further, the assertion regarding "the permanence of art" might have carried greater plausibility if the agency had not squandered its credibility by losing paintings or allowing them to be stolen.[3] Conspicuously, no mention is made of that contretemps. Also, according to the Senior Historian at the Bureau's Denver headquarters, Brit Storey, the collection has been scattered around the nation, from Boise, Idaho, to Washington, DC, a fact that makes it hard for interested parties to assess the permanence of the work. The reasoning about allowing "dialogue to occur" would be more plausible if any pundits had seen fit to parse the art or assess it in historical perspective before now; in fact, though, no one but Stinger and Ferguson seems to have given the paintings a second glance or wasted any praise. The rationale that "government-sponsored art programs seemed to be *en vogue* at the time" is true—NASA had chartered its own art program in the early 1960s—but it oversimplifies a complex issue, for NASA was never tried and convicted in the courts of public opinion in the same way the Bureau of Reclamation was (as will be discussed at some length below). Finally, though, the rationale that the artworks build upon "the tradition of western landscape painting" is perceptive enough to prompt elaboration here, inasmuch as government commissions fueled the Reclamation art program just as they did some of the most enduring western landscape art, a tradition that by now has been theorized enough to open a deeper analysis.

OVERTURES TO SUBLIMITY

In *Nature and Culture: American Landscape and Painting, 1825-1875*, Barbara Novak observes incisively that the nineteenth century's "apparently innocent nationalism, so mingled with moral and religious ideas, could survive into another century as an imperial iconography" (7). That observation rings true for the Bureau of Reclamation Art Collection, which arguably portrays an empire in iconographic terms. It builds an iconography of western public lands that have been "reclaimed" to advance and harmonize with Manifest Destiny. To make manifest the national fate to extend from sea to shining sea, the innocent nationalism that fueled expansion had to *take back* from nature, and from the indigenous people, lands that primordial forces long before had claimed. What preceded that reclamation—that taking back—would have been unacceptable to portray, even though there was narrative precedent in Thomas Cole's *Progress of Civilization* series. In that series Cole depicted humankind progressing from a savage state, to scattered communities, to a flourishing civilization, and down into a state of declension or decay where artifacts and memorials are laid waste.

One of the Reclamation paintings proves exceptionally reminiscent of the Hudson River principles that focus Hughes, Novak, and other commentators on nineteenth-century American art. It is *Olympus Dam*, by Xavier Gonzalez (see Fig 6.2). It recalls *Cotopaxi* by Frederic Edwin Church, whose trips to Ecuador in 1853 and 1857 resulted in paintings of that volcano in 1855 and 1862. In one of those pieces by Church, the mountain is erupting or smoking; in the other it lies dormant. The Olympus Dam, in the Gonzalez tempera, emerges from a wash of earth tones in the landscape, surmounted only by a snow-covered peak in the distance: Longs Peak in Colorado. In the painting it emerges like a peep of heaven, thus performing a trope on Greek mythology, in which Mount

Olympus is the home of the twelve gods. The dam—what one sees of it, scarcely visible amid the background shading into rocks and water—shares qualities of ancient architectural columns that further enforce the classical age. Those suggestive columns appear to be buttressing Longs Peak, as though the engineering feat were serving as a foundation for the mountain's grandeur. The painting elevates the dam to an altitude that rivals Longs Peak, exalting it above the spindly aspen saplings and common rocks in the foreground, through which an outlet stream flows. The painting exalts the dam as if the dam were a variation of paradise. In doing so, it envisions an imperial icon made from base concrete. The painting is an oblong lozenge, whose soft edges serve to soothe the tension of cultural conflict that occasioned the Bureau of Reclamation Art Collection.

In serving a healing or soothing function, the painting effectively reprises (minus any irony) "The Anecdote of the Jar" by American poet Wallace Stevens, a short poem from 1923 that is pertinent enough to quote here in its entirety.

> I placed a jar in Tennessee,
> And round it was upon a hill.
> It made the slovenly wilderness
> Surround that hill.

Figure 6.2. Xavier Gonzalez. *Olympus Dam*. Tempera. Olympus Dam, Colorado. Printed with permission by the Bureau of Reclamation.

The wilderness rose up to it,
And sprawled around, no longer wild.
The jar was round upon the ground
And tall and of a port in air.

It took dominion everywhere.
The jar was gray and bare.
It did not give of bird or bush,
Like nothing else in Tennessee.

The chief irony in this poem is the animated subservience of nature in the face of a technological artifact, the jar. Nature fairly genuflects. It pays obeisance to the force of a household canning jar. To suggest that a simple jar should have this power is improbable, hyperbolic, ironic. There is little evidence of irony in *Olympus Dam* by Xavier Gonzales. The actual Olympus Dam built by the Bureau of Reclamation forms Lake Estes, a major recreation reservoir that glimmers dully in the painting's middle ground. Dams are focal points in the landscapes they appropriate; like Stevens's jar, they are products of technology that dominate and constrain ecological processes. A technological artifact, the jar stands as a synecdoche for the entirety of those technological processes that gave rise to the jar, thus demonstrating in humankind a virtual rage for order. While the painting elevates the dam to a level that represents or at least suggests a higher moral order, the poem elevates the jar to an amoral supremacy above the land. The massive dam approaches divinity, while the minuscule jar supplants it.

The West has been subject to frequent assertions that it functions for the rest of the world as a resource colony, a far-flung part of a continental realm (see, e.g., Robbins 59–82). Bruce Babbitt, formerly Secretary of the Interior and Governor of Arizona, stated it this way: "Traditionally, the American West has been something of third-world economy based on resource extraction" (qtd. in Wilkinson ix). In that respect the public-relations makeover that the BuRec paintings aimed to help bring about—the pictorial recommendations of remade landscapes that they embody and imply—was not only a response to ideology. It was meant to quell more than strident environmentalist complaints. Broader public interests were involved. The economic fabric of western rural regions had been tattered by traditional conditions of cut and run, slash and burn, boom and bust. Federal interventions abetted those patterns, as it still does, via subsidies. In this respect the Bureau of Reclamation Art Collection aimed to effect more than simple damage control; it was reacting to more than the legislative power of environmentalism and the emerging scientific consensus that enabled it. Technology or applied science, especially federally subsidized technology, was implicated in negative models of failed sustainability. In his summary of these historical patterns of exploitation,

Richard Lowitt left the federal presence off the accountability list when he noted, "in the West, thanks to advanced technology, spectacular engineering feats, and ill-conceived agricultural practices, extractive enterprises of all kinds left in their wake devastated areas and declining communities that benefited briefly, if at all, from these corporate and metropolitan-based endeavors" (131). The BuRec, like corporations or metropolises, practiced unsustainable development under a guise of public works, and in that respect it may have served as a model for businesses and cities. Using the rallying cries of cheap power, flood control, recreation access, and economic development, the agency curried the cooperation of technically savvy corporations and western politicians, entities that sought in common to profit from the abundant natural resources and the federal largesse.

Only in retrospect—only since revisionist historians, novelists, and other intellectuals began to take the measure of these dynamics—has the federal presence come to be examined with much critical sophistication. Harmonizing with Bernard Devoto, Rick Bass, Edward Abbey, and Marc Reisner—to mention only the most vocal of the agency's critics—Wallace Stegner maintained of the Bureau that, "from the beginning, its aim has been not the preservation but the remaking—in effect the mining—of the West" (44). If Reclamation's aims were imperial, then, its water projects warranted celebrations laved with solemnity and pomp. Its projects needed to be dedicated, honored, and rendered iconic by artworks that might forever grace the nation and be displayed as trophies by the federal government to justify its expansionist energies and claims. Weighing on the government, too, was the noblesse oblige to patronize its working artists, and in so doing to inspire testimonials to the nation's greatness. High dams, arguably the grandest U.S. counterparts to Middle Eastern mosques and European cathedrals, came to serve quite well as monuments to empire.[4] The Grand Coulee Dam, built by BuRec during World War II, is the largest concrete structure in North America. Dam technology, then, perpetuated the tenets of Manifest Destiny and American exceptionalism.

Themes of American exceptionalism had pervaded the continental sense of self since John Winthrop preached his famous sermon aboard the *Arbella*, urging his fellow Puritans migrating with him to the Massachusetts Bay Colony in 1630 to see themselves and their experiment as conspicuous and exemplary, as "a city on a hill." Exceptionalism helps shape religions, based on faith that the Creator had selected the North American continent to be the culminating civilization in world history. If Nature had accorded American genius everything it needed to ennoble itself, then natural resources fittingly would be boundless and inexhaustible. Wide, fertile, this world without end warranted a correspondingly magnanimous amen, a generosity of the spirit, a national dream of development and growth. Paul Bunyan, Davy Crockett, John Henry, and Mike

Fink—demigods who tamed rivers, razed forests, and primed paths for civilization—were precursors to the BuRec's technology and the bureaucrats who applied the science to help dams come to pass. An exceptional people and land demanded a sublime art, one that made the perceptive painter working in the nineteenth century an amanuensis to deity, "God's stenographer," in the view of Robert Hughes. As Barbara Novak held, "Since the landscape was a holy text which revealed truth and also offered it for interpretation, artists who painted the landscape had a choice of what to transcribe and interpret" (7). That held true for commissioned painters as well; they, too, were free to interpret or read the landscape, though human technology and its outgrowths were apt to be determined most holy under the directive of a Reclamation patronage. The agent shaped the message, that is to say; the means foretold the end. The artists commissioned by the government were painting de facto landscape histories. "The overtures to sublimity in America's early history painting were readily transferred to the landscape, and [they] lead to a study of artistic rhetoric, that style of formal declamation which is the appropriate mode for public utterance" (Novak 19). To be held up as sublime is to be inspiring, in the etymological sense of God's respiration, God's breath, washing across the tender recipient. (In the sciences, of course, to sublime is to convert a solid to a gas without an intermediate liquid phase.) After the mid-twentieth century, after the gradual evaporation of God from the intellectual landscape, some Americans began to worship technocracy and its products in effect. Scientists, engineers, and technicians still enjoy high rank and political privilege.

Government art commissions began in the nineteenth century, and consequences of their work varied. Some artists played a part in exploration, some had a hand in conservation, and still others promoted tourism. A key critical distinction is whether the paintings resulting from those forays have lasting value or instead are documentary, that is, occasional (a discussion to be taken up in later pages). Just as the BuRec underwrote the paintings of its dam projects from 1968 to 1974, the Union Pacific Railroad underwrote the travels of landscape painter Thomas Moran on the journey of 1871 by Ferdinand Vandeveer Hayden. The railroad hoped Moran's work might help generate new tourist destinations. Instead, Hayden used Moran's watercolors as tools to lobby Congress for the protection of Yellowstone, which became the earliest national park, and thus prompted a belief that Moran himself had conservationist leanings. "*The Grand Cañon of the Yellowstone* was the first painting ever purchased by the U.S. government, having been acquired soon after the passage of the act making Yellowstone a national park" (Oravec 59). Such dynamic interchanges between conservation and expansionism are a neglected aspect of art history. As early as the 1850s, "well-known names begin to appear on the list of artists-explorers. John F. Kensett went west in 1854, and again in 1857, 1868, and 1870. Albert Bierstadt made the first of several journeys to the Rockies with Colonel Frederick Lander's expedition in 1859" (Novak 144). Thus the West became a proving ground for the earliest government patronage of visual arts on the North American continent. In 1853 Theodore Winthrop—two centuries after his Puritan ancestor, en route to Massachusetts Bay, preached his shipboard doctrine of American exceptionalism—traveled the Cascades and Northern Rockies. Following in the wake of a railroad survey, Winthrop noted in one of his books, "Wood-choppers had passed through the forest, like a tornado, making a broad belt of confusion" (70). A champion of progress like Winthrop, though, had to defend "the destruction that precedes reconstruction," or the devastation that technology so conveniently availed. Private rail lines soon afterward would transect bison calving grounds and transport dudes to the Great Plains to shoot the beasts from the safety of train decks. The BuRec sought similarly to make nature safe for humankind when it brought technology to bear on floods that altered wild rivers.

Such occasions for exercising novel technological control over nature would offer artists opportunities for western commissions. A biography of Thomas Moran analyzed his patron's part in legislative change this way:

> Hayden's prompt employment of an artist, once he could afford one, indicates that he and his superiors recognized the need for visual accompaniment to his written text. He understood that it was essential that he provide congressmen and other influential individuals with a literal view of the land he was covering, and his commitment to the idea contributed to his consistent ability to acquire greater appropriations than any other expedition leader. (Kinsey 49)[5]

Scenic art with a political purpose was a tool used by corporations and agencies alike, and congressional appropriations would remain a prize into the twentieth century. The goals of those appropriations differed from century to century, but the strategies were much the same. Especially when government commissions funded art, "sublimity as a convention functioned actively to screen, or project, human expectations and desires upon the landscape" (Oravec 61). As Andy Opel points out, the same holds true today in the commercial sphere for Yellowstone, where "public land is a stage upon which to project consumer desire" (111). The leading consumer desires in the twentieth century, as measured by agency decisions (or, some would argue, socially constructed by them), were access to wild lands, tourism, and motor-powered recreation. Just as the shrewd wise-use movement of the 1980s and 1990s adopted the language of conservation to set back conservation, the BuRec borrowed its visual-arts strategies from Hayden.

While the BuRec used art and artifice to manage its issues—to effect an "issues management"—its great dam projects themselves were forms of artifice. They artificially altered ecology and landforms by the devices of engineering. To reverence BuRec dam projects using the aesthetics of artistic sublimity, then, was tantamount to reverencing technocracy: a term used here to mean a philosophy that advocates the enlisting of a bureaucracy of highly trained engineers, scientists, or technicians to run society or the government. The only serious competitor with the Bureau for the title of top technocratic bureaucracy is the Army Corps of Engineers, and the competitive fervor that drove the two to build new dams furnishes much of the cynical entertainment value in Marc Reisner's *Cadillac Desert*. The new places on public lands—the dams, as well as the paintings that resulted from the attention lavished on dams—had to be acknowledged formally, had to be dedicated to confer monumental sweep. Formal acknowledgments and dedications served another function: to extinguish memories of and desires for pre-dam places. If the public had not laid eyes on a certain ecosystem prior to the incursion of a dam there, then expectation and desire in large part might be shaped by those artifices the public had beheld in museums, galleries, catalogues, and portfolios of the Reclamation Art Collection given away and sold in 1972, then given away and sold again in 2002 to commemorate the centennial of the Bureau's founding. Reproductions—virtual realities or virtualities, to coin a phrase—supplanted the original landscapes. In his introduction to *Uncommon Ground: Rethinking the Human Place in Nature*, William Cronon describes such exchanges this way: "Once we believe we know what nature *ought* to look like—once our vision of its ideal form becomes a moral or cultural imperative—we can remake it so completely that we become altogether indifferent or even hostile toward its prior condition" (40). And so the hopes of promoters have always inclined, whether in colonial times or in recent decades. Promoters aspired to redirect attention so fully—through language, music, or visual imagery—that that artifice or projection might displace the prior condition, the undeveloped or natural state.[6]

By these means the BuRec paintings were from their inception and remain today promotional instruments. Cronon wrote more to the purpose regarding the use of the visual arts in "Telling Tales on Canvas": "Booster images are ubiquitous in western art, so much so that they occur at every stage of the frontier process and often reflect promotional expectations far more accurately than they record historical landscapes" (44). For so many of the landscapes portrayed in BuRec-commissioned art, we have only that one portrait of that particular dammed place. There is no record of the place before the dam. Just as an artist's rendering of a resort or housing development helps to persuade potential clients or buyers, Reclamation artworks often idealize landscapes whose reality is more sterile, intrusive, institutional, or stark. To consider BuRec leaders in step with boosters and promoters is to see them in a newer light.

AN EMBATTLED AGENCY

The central figure in the ongoing culture battle that gave rise to the art program is "Floyd Dominy, the two-fisted commissioner who rode reclamation's falling star" (caption, 182ff). So wrote Marc Reisner, whose 1986 exposé of western water politics, *Cadillac Desert*, became a bestseller and was produced as a four-hour public television special by the same title. Commissioner of the Bureau of Reclamation from 1959 to 1969, Dominy oversaw some of the most ambitious and contentious projects in its century of ventures. Dominy's star began to fall as new environmental legislation made it more difficult for him to get congressional funding to impound big rivers. He rode the agency's star as it fell in the sense that he fought for the BuRec's credibility against arduous odds. By the date of the first Earth Day in 1970, the art program had been underway for two years, and Commissioner Dominy had been ousted for one. (Ironically he got his walking papers from James Watt who, as Secretary of Interior under Ronald Reagan later, would have an even more embattled tenure.) He retired to his cattle farm in Virginia, where he remained until his death in 2010. On the particulars of the art program, Dominy's fingerprints are nowhere to be found, but he could not have disapproved. His determined character (implicit in the noise of his name, like a character in a Restoration play) meant that he apologized to no one, least of all to conservationists, and he maintained great pride in his agency's work. As commissioner, he would have had to approve the commissions given to artists lured by the "international opportunity," and he would have had to approve attendant expenses such as helicopter tours, ground tours, river tours, personnel time to give the tours, and other promotional costs. The agency favored his and DeWitt's art program, for it continued four years after him. Dominy himself flourished much longer than that, and was still giving interviews in 2007 at the age of ninety-eight.

As the central character in an American culture battle, Dominy became larger than life, as respected by pro-dam forces as reviled by anti-dam interests. Environmental conflict has been slow to gain notice as a feature of the culture wars that got most recognition in the 1980s, but by any definition it qualifies. On the pro-dam side stand those who tend to favor economic development and growth, the harnessing of natural resources for the betterment of humankind, and the utilization of technological advances to yield enhanced qualities of life. In such arguments one may still hear whispers of Genesis 1:28 and its elemental exceptionalism: "God blessed them, and God said unto them, Be fruitful, and multiply, and replenish the earth, and subdue it: and have dominion over the fish of the sea, and over the fowl of the air, and over every living thing that moveth upon the earth." On the anti-dam side stand those who favor free-flowing rivers, muscle-powered sports, ecological integrity, and wilderness experiences that are not subject to the incursions of engines and turbines. Ecology, as used here, is well expressed by Shepard Krech III: it is to "think of the environment and its components in interrelating, systemic ways" (4). If the issues are complex, they often come down to one highly subjective measure: beauty.

Aesthetic measures often rise to the surface when partisans assess dams. If engineers and bureaucrats find dam designs attractive to behold, environmentalists in general think that they will never see a dam as lovely as a stream. The classic statement of the environmentalist position on this matter is Aldo Leopold's. "A thing is right," he wrote, "when it tends to preserve the integrity, stability, and beauty of the biotic community. It is wrong when it tends otherwise" (262). As Anne Marie Todd phrased it in *Environmental Ethics*, paraphrasing Leopold without acknowledging the debt, "An ecological aesthetic emphasizes the interconnectedness of ecosystems, linking nature's beauty to its ecological sustainability" (91). Todd also appealed to Aristotle to argue that perceptions often regulate ethics (90). Humans are sight animals, and the ways we see are culturally determined to greater or lesser degrees, or so some social scientists believe. In such philosophical contexts, then, the Reclamation art program may be seen as an agency effort to recover common ground, if not moral high ground, by claiming beauty for its water-development projects. Perhaps if it could regulate perceptions of its dams by placing its commissioned paintings before the public eye, then it could in turn regulate the ethical responses to those dams. In *The American Artist and Water Reclamation* catalogue, and in the touring exhibition that it accompanied, the agency endeavored to demonstrate by aesthetic means that dams, excavation sites, penstocks, spillways, batch plants, turbines, power-lines, trash-racks, irrigated fields, reservoirs, and feeder canals could be things of beauty. Such artwork constitutes an "aesthetic turn in green marketing," as Todd titles her article—an artistic innovation that gained government funding and succeeded in ways corporate and municipal organizations would emulate in years to come.

Several titles in American literature owe a debt of negative inspiration to Dominy's character and to the BuRec as it functioned under his control. Edward Abbey found literary fuel in his abhorrence of Glen Canyon Dam, whose gates opened in 1963, as did the radical environmental group Earth First! that Abbey helped to found in the desert Southwest. Jared Farmer, in his book *Glen Canyon Dammed: Inventing Lake Powell and the Canyon Country*, trotted through key events that led to later books by John McPhee and Marc Reisner. Farmer helped trace some historical, cultural, and literary legacies of Dominy's leadership.

> Edward Abbey floated the Glen in 1959, a once-in-a-lifetime experience that haunted him till his death. His novel *The Monkey Wrench Gang* is an extended fantasy about blowing up the dam. In 1981, when the founding members of Earth First! (the real Monkey Wrench Gang) went pub-

> lic, they chose the dam at Glen Canyon for their stage. From the lip of the monolith, they unfurled a three-hundred-foot ribbon of black polyethylene tapered at the bottom to resemble a crack (xiii).

The guerilla theater practiced there would bring thousands of adherents to the cause. Privately, Abbey used to refer to Lake Powell as "Dominy Falls" (Stiles). In 1968 he published *Desert Solitaire,* an irascible memoir that centers on his activities as a seasonal ranger in today's Arches National Park. In that memoir, too, he meditated on the act of blowing up the dam. He wrote in great distaste, "The impounded waters form an artificial lake named Powell, supposed to honor but actually to dishonor the memory, spirit, and vision of Major John Wesley Powell. . . . Where he and his brave men once lined the rapids and glided through silent canyons 2,000 feet deep[,] the motorboats now smoke and whine" (152). Abbey's objections concerned the reservoir's artificiality, its co-optation of the name of a man who should have condemned it, and its noise and air pollution. One scheme to sabotage the dam in *The Monkey Wrench Gang* involves a party barge or houseboat loaded with explosives that would detonate at the dam face. In Abbey's perceptions the dam clearly violated Leopold's ecological principles.

Another writer who reviled the agency, who carries academic credibility for having won a Pulitzer Prize, is Wallace Stegner. In essay after essay, and in the subplots of many of his novels, Stegner dramatized the crises of aridity in the West and the various ways he perceived the Bureau of Reclamation to have exacerbated those crises, including doling out federal pork to rural legislators. Stegner won a Pulitzer Prize for Fiction in 1972 and died in 1993. Harmonizing with Stegner, environmental studies professor Donald Snow still could write of the BuRec in 1997, "The greatest subsidy of all—measured by both economic and environmental impact—is probably the institutionalized giveaway of the public's water to irrigators and other users, who have spent the past century emptying the region's rivers and streams and converting as many as possible into reservoirs" (4). Snow's coeditor of the essay collection *The Next West,* economist John A. Baden, elaborated: "These projects continue to be built even though the economically and geographically viable locations for dams were long ago exhausted; and in spite of evidence that they harm many species of fish and valuable riparian areas" (123). Baden added: "the Bureau of Reclamation needs dams to preserve its budget and thus its employees' salaries and power." Such bureaucratic self-perpetuation suggests an exceptionalist ethos.[7] Stegner, a lifetime fan of public lands, which he capitalized as the Public Domain, clearly distinguished between the Bureau of Reclamation and other units within the Department of the Interior. He wrote, in *The American West as Living Space,* "In contrast to the principal water agency, the Bureau of Reclamation, which was a creation of the boosters and remains their creature, and whose prime purpose is technological conversion of the arid lands, the land-managing bureaus all have as at least part of their purpose the preservation of the West in a relatively natural, healthy, and sustainable condition" (38). The western public lands experiment continues to have its problems, including below-cost timber sales in the national forests, conflicts over grazing fees and allotments, and tolerance of hunting in some national wildlife refuges.

The book *Cadillac Desert* qualifies for inclusion in this growing literary backlash canon, so to speak. The book furnished powerful confirmation of much western American literature, whose plots find frequent form in conflicts over water rights, property disputes involving agriculture, land grabs, and the closing of the frontier. Dominy's efforts to shape public opinion had a long foreground, Reisner showed. Reacting to a story in *Audubon* magazine, for instance, the Commissioner "wrote to his regional director, 'We think it would be opportune and worthwhile to work with the Sacramento newspaper in the development of a feature story on the lengths to which Reclamation has gone . . . to enhance the fishery and wildlife resources of the Central Valley" (245). In sum, Dominy was pushing for a propaganda job in print. There is also in American literature the 1971 nonfiction classic *Encounters with the Archdruid* by John McPhee, perhaps the greatest living American writer working in the essay form today.

The archdruid of McPhee's book is David Brower: embattled general of anti-dam forces in the early decades of the conservation movement, erstwhile Sierra Club leader, later founder of Friends of the Earth. Part of the success of McPhee's book is its timeliness, for both Brower and Dominy were being booted from the leadership of their respective organizations, the Sierra Club and the Bureau of Reclamation, when McPhee conducted his interviews in 1969. McPhee arranged for Brower and Dominy to spend a week together on a river, and he records the strong opinions that alternately simmer and boil as the men debate. Long a writer for *The New Yorker*, McPhee pitches environmental conflicts in neutral terms. In one sentence he can madden anti- and pro-dam forces alike: "The conservation movement is a mystical and religious force, and possibly the reaction to dams is so violent because rivers are the ultimate metaphors of existence, and dams destroy rivers" (159). Brower might have rebelled against the first half of the sentence for its imputations of paganism; Dominy, against the second half for imputing that his agency was a ruinous force. Describing Dominy's efforts to dam the Grand Canyon, an ongoing conflict between the two men, McPhee writes, "These Grand Canyon battles were the bitterest battles of all. The Bureau felt that Brower capitalized on literary hyperbole and the mystic name of the canyon" (166). Literary hyperbole, if that may be considered a fair assessment of Brower's strategy, was to be answered trope for trope by the Bureau in the 1960s, when it would enlist high-profile artists who depicted dams with a visual rhetoric that could accord the agency a certain intellectual cachet. Toward the end of Dominy's weeklong encounter with David Brower, McPhee reported on a speaker at the dedication of the newly built Glen Canyon Dam.

> "A conservationist is one who is content to stand still forever," said Raymond Nakai, of Window Rock, the head of the Navaho Tribal Council. "Major Powell would have approved of this lake. May it ever be brimmin' full." Brower remained silent, but was having difficulty doing so. It was not hard to guess his thoughts. Major Powell—explorer, surveyor, geographer—was not alive to say how he might feel, in English or in Navaho. (196)

Even the neutral McPhee could see the slick façade Dominy had put on the ceremony. McPhee, Abbey, Stegner, and Reisner (not to mention many less well-known writers) demonstrate collectively that the battles to save America's free-flowing rivers had implications for the literary arts, as for science, politics, and the law.

Dominy made personal forays into aesthetics-as-marketing when he wrote a pictorial tract in 1965 and funded the production of a short film in 1969, both of them tendentiously entitled *Lake Powell: Jewel of the Colorado*. Neither one holds up as an example of effective PR nearly as well as the series of paintings and drawings DeWitt would later spearhead. The twenty-eight-page *Jewel* tract contains thirty glorious color photos, along with sundry black-and-whites and maps of the manmade lake, which extends 186 miles into Utah from the dam itself in Arizona. The tract also contains instances of poetry and prose. Jared Farmer called the former "engineered poems," which—according to an internal memo from the Bureau of Reclamation—were meant "'to combat the corrosive influence'" of the commercially successful Sierra Club coffee-table books that focused on ecological losses in Glen Canyon and on threats to the Grand Canyon (149). A good example of those poems is a four-liner entitled "Lake Powell."

> To have a deep blue lake
> Where no lake was before
> Seems to bring man
> A little closer to God. (13)

The ambivalence of the final line must not be lost. On a literal level a lake that is raised—where once there was a profound canyon and a river—may bring one higher in altitude to the sky where God is said to reside. On a more subtle level, though, the might of a species that can tinker with nature and create such a lake approximates the powers of God; technocracy as a godlike attribute must not be undervalued. But the lake only *seems* to bring one closer to God, and seeming is no closer to being than the poem and film are to actual art. Another poem assumes the voice of the river itself. Where it ran "brick-red, mud-laden," it is grateful to be

"Now flowing clean and blue / Unmaimed" (2–3). Dams do not maim rivers, the river itself attests. Printed in 63,000 copies, the *Jewel* tract was given to lawmakers and foreign dignitaries and was investigated as illegal lobbying (Farmer 150–51). The film version of the tract, released four years later in 1969, is twenty-eight minutes long. It opens with the words, "A winding sheet of blue, stretching across parts of Utah and Arizona, is Lake Powell." (A good editor would have recognized that a winding sheet is a garment to wrap a corpse in before it is buried.) As the film advances, a fantasy unfolds slowly. A family of four camps and romps through the wonders that are Lake Powell. The original classical score, composed for the film, segues into jazzy riffs when footage pops onscreen of fish being dropped into Lake Powell from the air by floatplane. Very soon, a mounted Indian appears dressed in red shirt, white pants, and a white headscarf tied behind. The solemn voiceover reminds viewers that this is the "land of the Navaho." The Indian looks approvingly at the skiers being towed behind a boat below. "Into his land came Lake Powell, which he has woven into his pattern of ancient ways." He leans over on his horse from atop a precipice and watches the family (and he reappears at film's end as if to assure viewers he is still content with his ancestral homeland). When the imagery and language move to the dam, the narrator's voice grows sonorous and inspired. "Impressive in its silent strength, the dam, with its ten million tons of concrete, rises 710 feet from sandstone bedrock." While the narrator names the various jobs and jobbers, footage of the excavation makes it seem as if building a dam were a traditional occupation, as if custom and culture were involved. The film segues to spectacled men in thin black ties and short-sleeve dress shirts—engineers who start the turbines moving. Scenes of pleasant camping and beach frolics follow, with fishing, bikinis, and boating abundant. A Lee's Ferry history lesson includes a tour boat with "Wahweap Marina" painted on the side, then more bikinis, stringers of bass, filets frying. Inspirational music leads to abstract ripples in the water, paired with pictographs, sandstone patterns, and a history lesson on Wild West miners. The family crouches flat to drink from Lake Powell then engages in a water fight. The music score moves again from jaunty classical to meditative when the "wonder" of the Rainbow Bridge appears—a natural stone arch whose ready access had become a source of much contention. "Now easily accessible by powerboat," it was seen only by Indians before the dam. Lake Powell serves a healthy leveling function, it implies. The feeble and the obese, the young and old, the white and red alike may now view the bridge close up.

The red-shirted Indian of this 1969 promotional film is worth pausing for. He would appear also in the oil-on-canvas painting by Norman Rockwell, *Glen Canyon Dam* (see Fig 6.3). His location in the painting, above the dam, and his red shirt are unlikely to have been coincidental. In the Rockwell painting he appears with his family—wife on horseback,

Figure 6.3. Norman Rockwell. *Glen Canyon*. Oil on canvas. Glen Canyon Dam, Arizona.

son beside him, scrawny dog taking up the rear. Members of this ensemble—threatened with extinction, pushed to the continent's edge—peer across a dizzying expanse toward the dam face. Their body language is more akin to resignation or forced cohabitation than approval. The dog cringes; two raptors soar above the gulf. Life goes on, even in the shadow of massive technology. "Over five million cubic yards of concrete make up the dam and power plant," one History of Lake Powell website boasts; "—that's equal to enough to build a four-lane highway stretching from Phoenix, Arizona to Chicago, Illinois." Unlike the BuRec-produced film, where the dam has been "woven into his pattern of ancient ways," the Indians in the Rockwell painting seem hindered by the concrete monolith in their way. Nomadic people historically, they see their travel baffled, slowed, constricted. A related American painting, *The Last of the Race*, by Tompkins Matteson, also depicts an Indian family group, including a dog, looking down from atop a precipice. Again the elevation suggests a moral order that has grown irrelevant. In its subject's vantage point, too, the Rockwell piece bears comparison to John Mix Stanley's *Oregon City on the Willamette River* (ca. 1850), in which an Indian couple overlooks a broad valley that has grown many white homes. Rockwell was working in an identifiable tradition by positioning his Indian family just the way he did. In the essay "Telling Tales on Canvas," William Cronon recognized in such representations an "icon that is repeated over and over in nineteenth-century frontier imagery: Indians witnessing a landscape of progress that also marks the end of their familiar world" (70). Like James Earle Fraser's eighteen-foot plaster *The End of the Trail*, such pieces are apt to be read one way by the dominant culture and another by the marginalized one, either as respectful memorials to a valiant race of people, or as reminders of conquest and defeat for those people. Mediating between those two extremes may be a kind of "imperialist nostalgia," a yearning for bygone times that absolves participants from complicity and guilt in the process that displaced native people and species, a theory that originated with anthropologist Renato Rosaldo (69). Norman Rockwell, borrowing imagery from the BuRec film, brought a dose of cold reality to the upbeat acquiescence depicted by the agency. (That film might have been required viewing for artists commissioned to portray water reclamation, as it was for John McPhee when he interviewed Dominy in 1969 as he gathered material for *Encounters with the Archdruid*.) Still and all, Rockwell's Indians appear to be survivors—able to surmount the odds of conquest and persist to survive.

The Department of the Interior, the cabinet-level agency that is BuRec's parent, was no stranger to commissioning artists to help its causes. In May of 1941, the Interior Department paid a folksinger to write songs that promoted the building of dams, including Bonneville Dam and Grand Coulee Dam on the Columbia River in Washington State. Two leaders of Reclamation at the time, Harold Ickes and Mike Straus, "cooked up the idea of hiring Woody Guthrie as a 'research assistant' to write some songs in praise of the dams" (Reisner 160). Guthrie, of course, is most well-known today for "This Land is Your Land." Money was available to be made, supplying the Bureau with the tools of the arts—provided the art could be made plain enough that the public could understand. A working-class icon, Guthrie traveled the Northwest in a government car, chauffeur-driven, and composed twenty-six songs in thirty days. "Many of the songs were instantly forgettable, Bureau of Reclamation press releases dressed up in cornpone rhymes," wrote Blaine Harden. "But the best song, 'Roll On Columbia, Roll On,' succeeded in ways that the publicists back in Washington could never approach" (81–82).[8] The song has a patriotic air, appropriate to wartime emotion and to what Stegner has called "the lunchbucket and 'American initiative' arguments of resource exploiters."[9] The song is written to appeal to people who view themselves and their land as outstanding and exemplary. Today the refrain line of the chorus, "Roll on, Columbia, roll on," comes off as ironic. It is ironic now to read inasmuch as the Columbia River today is at best a series of lakes separated by dams. Guthrie celebrated it, though, for the way its "power is turning our darkness to dawn," generating cheap electricity.

As the paintings commissioned thirty years later would accomplish more subtly by nonverbal devices, the lyrics of this song tap into a Manifest Destiny theme: "Tom Jefferson's vision would not let him rest / An empire he saw in the Pacific Northwest / Sent Lewis and Clark and they did the rest." Washington State adopted "Roll On, Columbia" as its official folk song in 1987.

More than sixty years after Guthrie wrote his song, more than thirty years after the Bureau of Reclamation Art Collection began, the Interior Department still was using art to market its policies and its technologies, as Michael Kilian reported in the *Chicago Tribune*. "The Bush Administration calls it 'Consensus Building Through Art.' I call it bureaucracy as art." Kilian continued dryly,

> To be sure, with its many paintings of expensive weapons systems at work and heroically bemedaled generals and admirals, the Pentagon for years has fostered war as art. There are some exciting pictures of threshers and reapers over at the Agriculture Department that amount to a genre you might call farm implement art.
>
> But through Oct. 31, Interior is displaying in its grand ground-floor museum a huge mural that embraces as its subject every department and section in Interior's vast bureaucracy, including the Minerals Management Service and the Bureau of Reclamation.

The mural in five sections took just two days to complete in 2003. Instead of enlisting or commissioning known artists for this project, Interior drew upon forty staff members under the guidance of artist Laurie Marshall, who had previously created art for NASA and for the U.S. Army Corps of Engineers. Murals by committee were supplanting fine art by recognized talents.

In nearly every media and genre, the Bureau of Reclamation and its parent agency the Department of the Interior have enlisted willing workers in the arts to ply their trade. To countervail negative public opinions, conservative and liberal alike, the government has used tax dollars to soften taxpayers who otherwise might rise up and rebel against the use of their money to fund dubious plans. Much like corporations spending money to convince consumers to spend money, agencies spent tax dollars to convince taxpayers not to object to its spending. Such expenditures seem designed to dodge obligations to address real problems. Since the installation of its thirty-one dams in the Columbia River Basin, the federal government has spent some $3.3 billion on the recovery of endangered salmon species in that basin alone, according to Taxpayers for Common Sense, a group dedicated to cutting wasteful government spending and subsidies. Perhaps that money is being spent with greater sincerity and consequence, however, than if yet another series of artistic ventures into public relations were undertaken.

A CRITICAL FRAMEWORK

Having reviewed much of the background pertinent to the Bureau of Reclamation Art Collection, and having touched on several of the paintings, we may turn again to the particular artworks and adopt a critical tool known as ecocriticism. This deliberately diffuse approach to the arts and culture is tough to typify, for it is united not by a theory but by a focus: the environment. Simon C. Estok offers a definition. Ecocriticism's chief goal is "analyzing the function—thematic, artistic, social, historical, ideological, theoretical, or otherwise—of the natural environment, or aspects of it, represented in documents (literary or other) that contribute to material practices in material worlds" (16–17). Ecological criticism, in short, studies how texts reflect attitudes toward nature.

The semantic range of "texts" is wide. Texts may range from a volume of poems, to a film, to a set of government documents. Texts can be verbal, visual, cultural, or musical. Taking the BuRec artworks as texts, we must acknowledge an insurmountable difficulty in detaching them from their historical contexts. Those historical contexts include the epiphanies in environmental literacy that the American public was undergoing in the late 1960s and early 1970s, the culture battles that helped prompt the commissions given to the artists, the need for the Bureau of Reclamation to mitigate the reputation it was gaining, the vogue in government-sponsored art collections or patronage, and the willingness of Commissioner Dominy to spend agency funds to make public-relations gains. Taken together, the mass of related texts that comprise the Art Collection may be seen as one cultural text connected by thin membranes. Having established the artistic context fully, we may now more narrowly refine the critical focus.

That focus is best expressed by the neologism *ecopornography*. While it has yet to enter standard dictionaries, it is current in key academic communities today, particularly in communication studies, marketing, literary studies, visual arts, and environmental studies. It originated with adman-cum-author Gerry Mander, who used it first in the periodical *Communication and Arts* in 1972, a date that makes the inceptions of the Bureau of Reclamation Art Collection and the neologism coeval events. At first blush the concept may seem trendy, easy to dismiss as the latest flourish of political correctness, but it perfectly fills an empty language function and is too important to reject as simply flip. Language is dynamic, changing and adapting constantly to admit new words, and to widen or narrow the meanings or semantics of other words. At this early stage in its development, *ecoporn* carries a moral smack and a visual edge; it disparages the analogous exploitations of nature and the human body. The word suffers from some semantic noise, i.e., what "occurs when people attach different meanings to words or phrases, or when a communicator's message is poorly worded or presented" (Demers 216). Besides the major weight of negative connotations the concept of *ecoporn* carries, for instance, it also has to shoulder a divergent European meaning that is largely positive. Practitioners of European ecoporn are second-language speakers of English, and perhaps for that reason Fuck for Forest, founded in Norway and based now in Berlin, maintains a members-only website and purveys what it calls ecopornography to save trees.

Straight porn and ecoporn both tend to objectify for aesthetic pleasure, for audience approval, or for commercial gain. Ecoporn resembles human porn in that "it masks sordid agendas with illusions of beauty and perfection," wrote Bart Welling, paraphrasing Mander (54–55). In his chaper in a collection of scholarship on visual ecology, Welling asserts that ecopornographic images "work to conceal both the material circumstances of their creation by humans and whatever impact humans may have had on the landforms and animals they depict" (57). To propose to apply this concept to the Bureau of Reclamation Art Collection is to call upon the consensus of historians who assert that the pattern of rampant dam building in the decades after World War II was, in retrospect, ill advised and unsustainable. Whether the commissioned paintings masked a truly "sordid agenda" would require deeper research, particularly into archival materials, than this brief introduction to the collection may undertake. Whether many of the paintings worked to conceal human impacts on the environment is a question that may be addressed with greater certitude. The expansionist ethos was alive far into the twentieth century, even if the reverential temper found its focus no longer in godhead but instead in humanism and human technology.

Like human porn, ecoporn traffics in staged intimacies or ecstasies. In the visual media, it may employ provocative lighting, tricks with perspective, and close-up shots to enhance and tantalize. As it applies to the BuRec paintings, such staging is implicit in the language of Stinger and Ferguson. They write of Reclamation as the "sitter," a willing subject that the sundry artists primed and posed, prompted and represented selectively and subjectively. The root word of *ecopornography* has an enormously broad (and growing) semantic range. There is currently in usage today, for example, "the pornography of poverty," a phrase used by ad people and public-interest organizations around the globe.[10] There is *The Pornography of Meat*, a phrase used to study the relations between carnivory and male dominance (Adams). There is also "the pornography of war" (Engel). All such usages imply degrees of titillation and exploitation. Pornography itself has been defined variously, but it typically involves morally objectionable depictions of sexual domination and degradation. Applying a hermeneutics of suspicion like this one to a heterogeneous series of paintings necessitates a disclaimer; those that lend themselves most to ecopornographic interpretations have been chosen for purposes of illustration here. Additionally, it must be admitted that many of the artworks remain aesthetically pleasing in their own right, apart from any agency agenda, nor should analytical intention ever squelch the possibility of beauty. Unyoking these artworks from their cultural context, though, removes them from historical narrative and from scientific understanding alike. If Bureau of Reclamation artworks incompletely disclose ecological contexts, that failure of disclosure is due not to hidden government agendas exclusively but also to limitations of the media and in constraints of the commissions. Just as human pornographers would not find benefit in telling the histories of their subjects or portraying the indignities they suffer, paintings added to the permanent collection of the Bureau would not portray the ecological damage created by a dam, for such a portrayal would be counterproductive to the intended message. And as public art, commissioned by an agency, displayed in catalogues and websites, these artworks bear messages.

In media studies, ecoporn is closely allied with greenwashing, a word that enjoys wider currency. In *Advertising Annual* for 2004, for instance, Ellen Shapiro wrote: "The word 'greenwash,' the dictionary tells us, is formed from two existing words, 'green,' committed to the environment, and 'whitewash,' a deliberate attempt by concealment to clear the reputation of a person or an institution" (196). Corporations endeavor to cleanse themselves of culpability by making over their images in other hues. They camouflage themselves as a moth on hickory bark might. Such analogies, however, break down when one considers the active rather than the passive nature of greenwashing enterprises. Closely allied to such organizational practices is astroturf marketing, so named for the ersatz grassroots surface it displays. As Michael E. Kraft and Diana Wuertz note, it is "often called astroturf campaigning because it is said to create the illusion of public opposition, when in fact most of the letters and telephone calls are arranged by public relations firms employed by industry for this purpose" (119n2). Again, the walls between government strategies and tactics and corporate approaches totter when one critically examines the practices that the Bureau of Reclamation pioneered as early as the 1960s, practices that later came to be mainstreamed. Such practices would be less apt to succeed today due to the greater sophistication of viewers, their growing visual and environmental literacies. From greenwashing and astroturf campaigns, a protective coloration ensues, a green façade that makes it harder for detractors to single out the worst organizations for castigation or blame. Whether it is alluding to public relations, to art, to literature, or to landscape photography, ecopornography has pejorative connotations. Significantly, the seminal statements that most help quiet the semantic noise of *greenwashing* and *ecopornography* originated from publications in the advertising industry.

READING THE ART COLLECTION

To return to the BuRec collection, it would be well to recall the paintings by Kingman, Gonzalez, and Rockwell. All three depict dams. *Flaming Gorge Overlook* by Dong Kingman makes the dam a site where tourists assemble; they paint, play, and recreate. The dam resembles a giant easel, its concrete edifice a place to project human desires and expectations vis-à-vis the landscape. Films and laser light shows, in fact, long have been projected using dams as screens. At the massive Grand Coulee Dam, which Woody Guthrie immortalized in song, people come from miles around to take in colorful shows. Both the BuRec and the Army Corps of Engineers spend lavishly to promote tourism. The painter Dong Kingman was exaggerating, but not by much, when he sketched in more than a dozen walkers, bikers, painters, and photographers in his watercolor. Like Cole's famous oxbow—a ready emblem of restraint and control—the dam drawn by Kingman brings nature into submission. Much as a purveyor of human porn might ask the subject of his portrait to feign passivity, Kingman made the sedate environment of his dam appear inviting and docile. The lacunae of species indigenous to the region, the tons of water, and the held-back current—all have been made to submit to human economics and will. Lilliputian humans crawl across the body of the subdued beast. The *Olympus Dam* tempera by Xavier Gonzalez likewise makes the dam an object of aesthetic pleasure, one that is ineffable, ethereal, detached, so that the dam appears to have no connection to the massive granite blocks in the foreground. As if it were floating on clouds, the dam is *in* this world but not *of* it, an artifice, an ideal devoutly to be wished for. Hydrology and other sciences tell us, however, that all dams accumulate silt, which backs up behind the dams and in doing so suspends key ecological processes. This was a point of sore contention between Brower and Dominy, as it is for their heirs today. In time, a given dam might or might not fill up with silt and no longer function properly. In short, heavy geological facts belie the lofty, airy, delicate quality of the Gonzalez picture. Captions in the catalogue admit as much, though to gain an ecological understanding one has to read them backwards, so to speak. The caption for the watercolor *River Bottom* by Lee Weiss offers an example: "The Green River in northern Utah, murky and opaque before construction of Flaming Gorge Dam, now runs clear over the rocky streambed."[11] Such captions privilege the visual and ignore the ecological fact that the dam under portrayal now is gathering and capturing silt that the current used to carry downstream, silt that once was deposited as rich alluvium. As for the Rockwell piece, it is one of few extant to include humans in its depiction of a BuRec dam site, and it is the only piece to feature Indians in *The American Artist and Water Reclamation* exhibition from 1972.

Rockwell's *Glen Canyon Dam* canvas forces viewers to see the dam from the Indian perspective, both physically and philosophically. The mother earth revered in Indian animism appears to have been defiled, her canyon blasted and carved, her life-giving waters reduced to a trickle. The dam is a tourniquet, a technological constraint that any fish remaining downstream will not get past. From the Indian perspective, the engineering achievement may be considered to be manipulative and abusive, a sacrilege, a scar that will necessitate decades to heal. Still, the painting gives every indication that the Indians' resolution to survive will serve them in good stead. Pacific Northwest residents who live near the once-mighty Columbia River watched during the spring of 2000, a high-water year, as Indians gathered in north-central Washington below Kettle Falls, gathered and chanted, hoping the increased flows were signaling a return of the salmon runs the Grand Coulee Dam had blocked more than sixty years before.

Ecocriticism invites alternative readings, perhaps from the perspective of indigenous peoples, perhaps from the angle of science and its methodologies. Nothing should have compelled the BuRec-commissioned artists to paint what lurked below the surfaces of those dam projects. Inasmuch as agents are apt to shape the content of agency-commissioned artwork, and insofar as means tend to foretell ends, then any art that offered negative commentary in the process of depicting the "imaginative aspects of the Reclamation Program" would almost certainly have been excluded from the exhibition and catalogue in 1972, if not also from the mismanaged permanent collection of some 355 or more pieces.

Lake Powell, due to its high profile in the 1960s-era pile-up that damaged the Bureau of Reclamation, comes in for an unusual number of artworks in the collection. *Campsite at Dawn*, an oil-on-panel piece by Dean Fausett, strives to show just how placid and serene the reservoir named Lake Powell could be (see Fig 6.4). As the title suggests—and as liner notes included on a poster of the piece confirm—the painting depicts a site where the artist camped as part of his commission with the agency. That site is near the location where the Escalante River enters Lake Powell, a river that would itself become a flashpoint in later battles between the state of Utah and the federal government after President Clinton, in one of his last acts in office, protected or set aside miles of public lands as a national monument. In the close foreground of the painting, a half-dozen stalks of yucca bloom. Their tall, white flowers, ascending in clusters up the plant stems, resemble smoke in the middle ground, where tents are pitched and boats pulled up. Most remarkable about this painting is its use of light. On the water of Lake Powell, on the cliffs in the distance, and on the wispy clouds, the early-morning glow appears apocalyptic. Exceptionalism is implicit in the scene. As if God had parted the clouds to favor the camps and the dam, a space in the sky above the campsite deepens to an inky blue. Using Hudson River aesthet-

Figure 6.4. Dean Fausett. *Campsite at Dawn*. Oil on panel. Lake Powell, Utah. Printed with permission by the Bureau of Reclamation.

ics in its landscape scale and its lighting, the piece resembles an overexposed photo, too flooded with a too-stark radiance, as if illumined by a nuclear explosion. The state of Utah had grown infamous for its nuclear tests in the 1950s; residents used to settle in and watch the tests for entertainment. Terry Tempest Williams used those tests as backstory in her book *Refuge*. That book expresses her belief that her grandmother and mother contracted breast cancer from the tests, prompting her to characterize herself as a member of the "clan of the one-breasted women." The clouds in the painting seem blasted or blown, though there are no ripples on the lake. The liner notes, on the poster for the painting, slip in a none-too-subtle PR plug: "It is typical of the magnificent scenery which is attracting recreation-seekers to Lake Powell from all over the United States." Overlays of language and paint drive home the PR message hard.

Some paintings from the Bureau of Reclamation Art Collection collapse the twin meanings of ecopornography into one. *Sealing the Crevice* by Fletcher Martin represents (the caption on the website says) the act of "sealing a fault in the granite wall of the forebay reservoir for the Grand Coulee Third powerplant, Columbia Basin Project" (see Fig 6.5). But the painting is more complicated than the engineering sketch at first suggests. On a level of pure imagery, the crack is vaginal, it must be said, just as the adjuncts of the crevice have been given the shape of legs. Whether the actual fissure or crevice at Grand Coulee Dam looked the same is immaterial; the fact is that the watercolorist cast his geological subject matter in terms of human anatomy. Compositionally speaking, the legs funnel the viewer's vision down to the anus, just as gravity channels

Figure 6.5. Fletcher Martin. *Sealing the Crevice*. Watercolor on paper. Grand Coulee Dam, Washington. Printed with permission by the Bureau of Reclamation.

water down to deepen and enlarge a stone gap. Not in invitation but in voyeuristic prurience the eye is made to follow the line. But this watercolor painting offers another perspective as well. To regard the painting from the angle of vision of the men who are engaged in sealing the crevice—an angle that the piece invites—necessitates an eye-level view of the sexual crevice. From their angle the gaze trends upward along the basalt legs that rest clinically in the air, as if the feet (off the frame) had been positioned in stirrups in a gynecologist's chair. Instead of the tools a gynecologist might use, a crane, a truck, and a massive plug are sealing the crevice or crack. What coloring there is in the painting lends itself to an anatomical understanding. The striations on the crevice walls imply the muscles in a set of thighs and buttocks, while the crevice in red suggests flesh. Or perhaps that red at the center suggests the bloody mysteries of menstruation, one of the most complex of subjects in male art. To speculate why the crevice (crack, fissure, fault) should need to be sealed would plunge one interpretively into the work of Freud, Lacan, and psychological scholars and critics. Not to lose sight of the literal and geological, I explore them below.

The present reading of the painting leads to intimations of fecundity and sexuality obstructed. The plug that seals the crevice ensures no leaks. The earth mother is being unsexed, sealed up, and forcibly made chaste. Nature forcibly has been brought under technocratic control. No more seepage will hamper the dam, no more betrayals or stains. The flawed rock again has been made whole. In this respect, *Sealing the Crevice* bears intimate comparison to *Mother Earth Laid Bare*, by Alexandre Hogue. The difference between the two paintings is that the Hogue piece is allegorical and thus leaves little to the imagination. The artist commissioned by the BuRec, Fletcher Martin, might not have consciously been adding human anatomy to his piece, for which he chose a title that includes the word *crevice*. In geological terms, crevices may compound the geological problems associated with "bank storage absorption of water into porous rock" (Nash 108). The word *fault*, though, was chosen by the one who captioned the painting on the Bureau website. With all its moral baggage, that word may be alluding to an unspoken quality of wrongdoing, blameworthiness. Remembering the Latin *culpa*, one realizes that a sense of culpability has been pressed upon the picture. A feminine environment is at fault, has fallen, and a stern and sturdy technocracy must generate the best corrective.

Nothing is unique in a pattern of assigning the feminine gender to nature. It is one of the oldest tropes in the history of humankind, going back at least as far as Greek mythology, wherein father sky pressed himself on mother earth and thus conceived the cosmos. There was to be found at the artists' disposal a ready storehouse of images and analogies from which "to depict the imaginative aspects of the Reclamation program." The language of the original ad for artists again is significant here; it must have liberated them in ways that other government commissions possibly curtailed. Far easier to have dissolved into abstraction, as did Richard Diebenkorn and Lamar Dodd in their pieces for the program, than to have kept grounded in the representational while still granting special latitude to the mind's eye.

The catalogue of *The American Artist and Water Reclamation* contains language that subtly belittles those who question the environmental legacy of the Reclamation Bureau. Douglas MacAgy of the Smithsonian Institution wrote the introduction. In it, he located the BuRec artworks in frontier pioneer days. "Today's review is staged before the same awesome backdrop," he wrote, "after generations of its mutual adjustment to human need and will" (5). The landscape and the ecology, that is, have had to undergo a mere "adjustment" that leaves the "awesome backdrop" the same. Like a human back out of alignment, the land under the purview of Reclamation needed to be adjusted, handled, manipulated to be brought back into working order. "Viewers," he continued, "now bring with them some notions about technics, economics and ecology—abstract terms often worn thin by glib usage until their users meet the facts face-to-face." MacAgy's language is difficult to unpack, but the condescending tone is plain. The little learning of the dam detractors is a dangerous thing. They are glib—fluent and smooth. They toss around terms that they don't understand, but those terms crumble once one confronts the pictorial evidence of how much beauty remains. Again, visual considerations take precedent over everything else. Condescending, he suggests that even a vicarious experience of those landscapes may offset everything that technology, economics, and ecology tell us. His faith resides within the tyranny of the eye. "[I]ndividuals who stubbornly hold on to a feeling for wilderness" "don't contest the importance of technology, but their investment of sentiment in the virginal aspect of nature is not about to be liquidated." That is, stubborn souls who feel fondly toward nature in a pristine state will not relinquish or liquidate that "sentiment," a use of the word that diminishes concerns about health implied by ecological concerns. All such concerns are reducible to "the quality and degree of sentiment that survives romantic attachment to a symbol." Such people, stubborn and idealistic, are out of date and should not be heeded, because "the lingering romance in their sentiments about picturesque nature is in many respects obsolescent." That is, they are behind the times. Likewise passé is literary fiction about the West, he says outright, for "the scene of real adventure has shifted from wagon trails and shack towns to sites of impressive technological innovation." Duly impressed, thoughtful citizens "have been caught up in the excitement of man's grandiose teamwork with nature." As a foursquare proponent of the visual arts, MacAgy grants entire primacy to visual responses to the landscape. In doing so he ignores, belittles, and embargoes the health of ecosystems as measured by scientific research.

CONCLUSION

On a ten-week professional leave to study ecopornography in 2007, I found my focus returning again and again to the Bureau of Reclamation Art Collection as the best instance of ecoporn. That art collection, described and interpreted here, began in 1968 as a way to deflect public attention away from the harmful effects of western dams and canals. The Bureau commissioned forty painters to explore the allegedly overlooked dimensions of Reclamation projects in fine-arts media. Some of the paintings remain memorable for genuine beauty and originality, while others are more blatantly promotional in thrust. The aesthetic challenge those painters were asked to address, "to depict the imaginative aspects" of Bureau of Reclamation dam sites, was badly worded or misleading. In point of fact, the artists were being asked to project their imaginations upon the dams, the dam sites, and other reclamation projects. That the resulting paintings should contain discernible subtexts of empire building, dominion, and subjugation was almost inevitable from the start. The trick the painters had to pull off was to accommodate the agency's need at the same time they fulfilled the mandates of honest art to address some of the environmental implications of the dams.

If the leaders of the Bureau of Reclamation had hoped the agency would be able to draw upon tax dollars to propagandize dam sites and easily influence public opinion in their favor, those bureaucrats were to discover that the cubic feet per second of cultural change could not be redirected. Like a small boat capsized in turbulent water, the agency would not again regain enough buoyancy for its grand dam-building spree of the previous decades to continue as before. Not only had Americans seen through the agency's technological agenda, but most of the viable dam sites had also been already claimed and developed. In the once-great Columbia River watershed where I live, the only free-flowing stretch of water, the Hanford Reach, has retained its status as salmon spawning habitat only by virtue of its high security as a storehouse of nuclear materials and an erstwhile site for manufacturing atomic bombs used in World War II. In an irony of history and ecology, those public lands and waters reserved for purposes of munitions testing and development, for technological innovation and design, like the Hanford Nuclear Reservation in Washington State and the White Sands Missile Range in New Mexico, remain some of the only places still functional as habitat. For its free-flowing water and relatively pristine state, the Hanford Reach is more pleasing visually, and more sustainable ecologically, than any of the hundreds of dam sites the Bureau of Reclamation developed.

NOTES

1. Consider this 1971 summary, from *CQ Researcher*, of rising levels of activist engagement and their impact on the Bureau: "Conservation groups actively pushed for many of the protective measures Congress has enacted over the past decade. They have also lobbied, often successfully, against bills they decried as harmful to conservation interests, and on occasion have gone to court to block action by some arm of government. David Brower, then the executive director of the Sierra Club, is credited with bringing together in the 1950s a coordinated lobbying effort by conservationists to prevent the Bureau of Reclamation from building a dam at the juncture of the Green and Yampa rivers that would have flooded sections of Dinosaur National Monument on the Utah-Colorado border" (Blumenfeld).

2. "The history of public art is enormous and includes commissioned art as well as art initiated by artists," writes Marc Pally for the National Endowment for the Arts. "Planning efforts for public art are, not surprisingly, almost exclusively directed toward official activities: government sponsored programs that either depend on mandated allocations (e.g., one-percent of capital improvement project budgets or private development projects), or that encourage voluntary participation through incentive programs or through the power of persuasion and mutual benefit."

3. See, for example, "Valuable art missing from Interior Dept." *San Francisco Chronicle* Aug. 18, 1990: "Among the [missing] pieces are a Norman Rockwell drawing of the natural rock formation known as the Rainbow Bridge, and scores of paintings and sketches commissioned by the department during the past few decades. In all, at least 357 artworks are missing, the investigators said." "Department investigators said the missing works are worth millions of dollars, although it is impossible to place a precise value on them because of inadequate records." "[Department Inspector General James R.] Richards said the most disturbing loss occurred in the federal Bureau of Reclamation, a unit of the Interior Department, which had been unable to account for nearly half of 355 paintings it commissioned in the 1960s and 1970s." Many major newspapers in the nation covered this story when it broke in 1990, and two years later the Bureau claimed to have "strengthened its accountability and controls over the remaining 201 pieces of art" ("Management of Artwork").

4. Attention to dam design may be found in Condit, Wilson, and Moffett and Wodehouse.

5. See also Peter Boag, who notes, "In 1936 the National Park Service described Moran as the 'Father of the National Park System'" (42). Boag refutes the notion, though, of Moran's conservationist leanings, arguing that the overtures to sublimity in Moran's work had roots in personal, rather than political or conservationist, sentiments and experiences at the time.

6. Joni Louise Kinsey offers corroboration: "'After seeing [Moran's] paintings, visitors to the West often were unable to view the sites without preconceptions; the images forever changed perceptions of the land and its meaning for the culture that inhabited it'" (qtd. in Oravec 59).

7. On the ways political institutions enforce exceptionalism, Paul W. Kahn wrote, "The final cause of the modern nation-state is nothing more than the perpetuation of the state's own existence. The state is not trying to achieve any end apart from the continued realization of itself. In it, means and ends collapse into each other. It is a means to no other end than its own existence. Thus the American nation-state understands itself as a temporal project without limit" (214). The Bureau of Reclamation also collapsed means and ends, soliciting known-quantity artists who would tractably enhance the public relations it so sorely needed.

8. See also Paul C. Pitzer, who dubbed them the "small army of Federal publicists" and the "beggars of superlatives" who promoted the dam (164). The BuRec managed also to confederate with journalists and with the Northern Pacific Railroad, "which included in its timetable the comment that those traveling in the Pacific Northwest

should consider a side trip to the dam" (166). "By the start of the 1936 tour season, the Bureau of Reclamation had built grandstands at vista points on each side of the river."

9. Wallace Stegner, "It All Began with Conservation." A fuller context for this quotation, in the conclusion to Stegner's *Smithsonian* article, is as follows: "And so this is the long road the nation traveled to get to Earth Day 1970. Those who led us down it, the countless individuals and scores of organizations who resisted the lunchbucket and 'American initiative' arguments of resource exploiters, did not leave us a bad legacy."

10. See, for example, British Overseas NGOs for Development, which used the phrase in the following way: "Though disaster relief and child sponsorship messaging still tends towards pornography of poverty stereotypes, most NGOs now avoid overly negative images and messages, replacing pictures of desperate people with pictures of smiling women and children, for example. Yet photos and stories often still lack a sense of context – they might represent people in need but do not show the obstacles that those people are working to overcome" (BOND). The locution "messaging" places the discussion in public relations.

11. A very similar piece, *The Green River*, also by Lee Weiss, bears a nearly identical caption: "The Green River, murky and opaque before construction of Flaming Gorge Dam, now runs clear over the rocky bottom" (17).

SEVEN

American Nature Writing and the Wise-Use Movement

The counter-environmental wise-use movement impedes the conservation of species and ecosystems, as a number of books attest, but it also obliges scholars to modify their definitions of nature writing. Traditional nature writers, as Scott Slovic has noticed, vacillate "between rhapsody and detachment, between aesthetic celebration and scientific explanation" (4). Yet some nature writing can be quite militant, like the Edward Abbey best-sellers *Desert Solitaire* and *The Monkey Wrench Gang*, which cultivate indignation and outrage by integrating political and scientific arguments. And all forms of environmental writing arguably construct nature, as Neil Evernden has demonstrated, by constructing our perceptions of it.[1] So it is with the burgeoning literature of the environmental backlash, a literature that sets out to correct a perceived disparity between economic growth and nature, between human rights and nature's rights. When advocates for nature deploy the science of ecology, as Abbey and other activist writers often do, they often set out to contest American industry and technocracy whose profits hang in the balance. They challenge the status quo. The rise of backlash literature has been understood as industry's answer to the assertions of ecology, to the growing popularity of nature writing, and to a series of scientific conclusions that stymie industrial progress. To say it another way, the counter-environmental wise-use movement (WUM) offers antitheses to the theses that drive nature writing and environmentalism. What I call wise use counter-science (non-consensual scientific judgments promulgated by organizations whose goals are political, ideological, or industrial) challenges the legitimacy of scientists and nature writers alike. Frederick Buell dubbed such writers and activists the "ideological fire brigade" (xvii). Like so many of their rivals in the environmental movement, writers who side with the WUM socially create nature in their own image.

The canon of American nature writing and environmental literature is broader than conventional wisdom allows. It encompasses more than scientific assessments and rhapsodic passages. It also often entails an implicit or explicit culture criticism. Indulge me in a very brief example of such culture criticism through a nonfiction classic of American nature writing. Mary Austin's *The Land of Little Rain* (1903) is a book appreciated for being rhapsodic. Set in California's Owens River valley, the book profiles a land and its humans, including "The Pocket Hunter," an itinerant miner who strikes gold and heads to London to spend it, where "the English middle class, with just enough gentility above to aspire to, and sufficient smaller fry to bully and patronize, appealed to his imagination, though of course he did not put it so crudely as that" (24). What do class issues have to do with nature? The land, Austin suggests, attracted the man and gave rise to his austerity of character, for dry landscapes ordain certain behaviors. This notion increasingly is coming to be known as geographical determinism.[2] The intersection of culture and landscape is more overt where Austin issues religion an ecological reality check. "'Come,' say the churches of the valleys, after a season of dry years, 'let us pray for rain.' They would do better to plant more trees" (52). Along with class commentary and religious commentary, Austin's book implicitly reproaches sheep and cattle ranchers. "Now and then some [pine] seedling escapes the devastating sheep a rod or two downstream" (37). Critiques of our cherished cultural institutions, combined with fresh environmental observations, lie at the heart of this book and of other classic American nature writing. Rhapsody is more correctly an accouterment, not a central feature. Even Annie Dillard, perhaps the most rhapsodic of contemporary nature writers (whom Robert McFarlane has singled out for her "zig-zag-zaniness"), criticizes our and her own obliviousness to creation's wonders, a criticism that lies at the center of her Pulitzer Prize-winning *Pilgrim at Tinker Creek*. The twin prongs therefore of my broadened definition of American nature writing are that it builds upon a base of culture criticism and that it focuses predictably on nature as the site of that criticism. I argue to expand the canon so that students of American civilization may understand, appreciate, and confront the burgeoning anti-environmental wise-use movement with greater intelligence and care.

NATURE'S RIGHTS

At the University of Idaho in 1993 I was moonlighting as the coordinator of a business communication laboratory while at the same time teaching a load of literature courses. In my spare time I was dreaming about having the autonomy to design a class that would navigate the nexus between ecology and culture. In that College of Business I was a rare bird who liked to read poetry and stories, so the business professors and dean liked to engage me in ideological debates. They endeavored to pick my brain, puncture my pretensions, and see what made me tick. One day I was toting around a book, British critic Jonathan Bate's groundbreaking study that had just come out that year, *Romantic Ecology: Wordsworth and the Environmental Tradition*. A professor of management asked me to characterize it. In terms that I recognize now as dangerously simpleminded, I summarized that book's thesis: the poet Wordsworth began as a socialist who cared for the common folk, but as he matured, his circle of concerns grew to encompass other species. The socialist became an environmentalist. "Socialists and environmentalists!" that University of Idaho management professor thundered. "Not a lot of difference in my mind between the two. Both of them try to tell me what to do with my private property."

Especially in resource-dependent states like Idaho, where I taught for a total of seven years, first at Idaho State University in Pocatello, later at the flagship institution in Moscow, *socialist* and *environmentalist* are hot-button terms. It took me time to learn how people confederate them mentally. Members of the wise-use movement (as they denominate themselves) like to name such covert socialists "watermelons," that is, green on the outside and red within. Like Native Americans who have forgotten their heritage (their critics call them "apples") and over-assimilated African Americans ("Oreos"), environmentalists who have socialist agendas are construed as fruits that strive to cloak their true hues. Many American leaders of business and industry, along with Tea Party senators, representatives, and county commissioners in my area, define environmentalism as a deadly subterfuge for the real agenda of grabbing hard-won wealth. Likewise, Christian patriots sometimes fear environmental ideologies as covertly encoded paganism designed to brainwash children into worshipping nature. To stake out some of the parameters in these culture clashes, to investigate how they bear on the environment and ecocriticism, indulge me as I survey some of the beliefs that environmentalists share.

Legally and philosophically, most greens agree that the natural rights assured by the Magna Carta and the Declaration of Independence ought to extend to certain parts of nature. Indeed, a number of our laws already do. We may not beat or starve or mutilate a domestic animal without fear of legal reprisal; may not take motor vehicles in wilderness areas since the passage of the Wilderness Act of 1964; may not legally harm threatened creatures, not since Richard M. Nixon signed the Endangered Species Act in 1973. Such laws were made to protect nature, to extend it decided rights; we depend on nature, the argument often goes, and when we do it harm we jeopardize ourselves. Peregrine falcons are recovering nicely since we stopped shooting them and adding DDT to their diets. One of the proverbial canaries in the coalmine, the falcon warned that portions of our planet were imperiled. All aside from selfish motives and grounds, visionaries as diverse as Saint Francis and Albert Schweitzer have argued that it is ethically wrong to base nature's worth on our own needs. For present purposes, we may define ethics as "extended sympathies" (Nash 45). Extending sympathies to nature helps create the "expanded community on which environmental ethics rests" (20). In short, ethical standing no longer begins and ends with human beings, just as we no longer restrict our ethical regard to propertied white Christian males. We look back with astonishment at a time when such boundaries applied. In a few generations we might look back with as much amazement to a time when we excluded rivers, oceans, and other species from our ethical circles. Meantime, cultural clashes aplenty are taking place. As we extend rights to nature, human "custom and culture" suffers, or so the WUM argument goes. Environmentalism becomes politicized and occasions a backlash, much of that backlash dedicated to arguments that say our uses are in fact sustainable and nature is in no clear and present danger.

Views on the rights of nature need to be further excavated. Because they have faced a climate of hostile opinions for centuries, they "took what shelter they could find in the humanities, in religion, and in the so-called counter-culture" (Nash 73–74), a remark that supports my revisionist definition of nature writing to embody a culture criticism. This remark also explains how English professors come to environmental issues. To read literature ecocritically, from an ecological perspective, is to excavate ideas that found little acceptance in our ethics, politics, and laws until the late twentieth century. Another reason readers might want to discover nature writing is to reclaim regionalism. Some of the most vivid stories and poems spring from intimate familiarity with backwater locales, with the tulles and the boondocks. For generations professors have puffed about how the best literature should be universal. Regionalism, along with its retarded cousin local color, "reeked of provincialism" (Parini 60). Who wants second best? Reading ecocritically, Americans now understand Lame Deer and Black Elk as culture critics with environmental messages. Indeed, we can read all Native American literature anew for its green views. Now we can discover riches in dozens of yet-to-be-remembered regional writers. I am talking here about a "placed" literature as opposed to a placeless or interior literature. Edith Wharton's fiction, for instance, however masterfully nuanced, is mostly located inside her characters' minds, which is not to say she may not be read ecocritically. Literature of the environment asks us instead to discover, as John Elder has said it, "some larger grammar in which the words *culture* and *wilderness* might both be spoken'" (25).

In the following visual aid, comprised of three intersecting circles, I try to codify the elements of environmental literature. Traditional nature writing is the top circle, built on rhapsody and aesthetic celebration. The human endeavors of criticism and science intersect with nature as integral parts, molecules if you will, of this literature. Science is integral insofar as it is a theoretical basis of knowledge about the natural world, which in turn provides us life and a focus for our art. Criticism is a theoretical basis also, a basis for understanding human worlds. By suggesting that criticism intersects with nature and science, I mean to insist again that writers of environmental literature typically set about delivering an implicit or explicit cultural criticism. They utilize nature as a way to critique human endeavors. Thoreau's *Walden*, to my eyes, is not a book primarily about nature. It is a book that criticizes human foibles and institutions vis-à-vis nature. Edward Abbey and Thoreau abide in that nexus between nature and criticism. Between science and criticism, I position as examples E. O. Wilson and Jared Diamond, Pulitzer-winning writers of non-fiction who wield science (in which they are accomplished or finished) as a tool of criticism. Dillard and Loren Eiseley, practicing an "ecstatic science" (Elder 3), emerge from a fusion of nature and science.

My examples draw chiefly from non-fiction, which seems to be the favorite genre of nature writers—from early American natural historians,

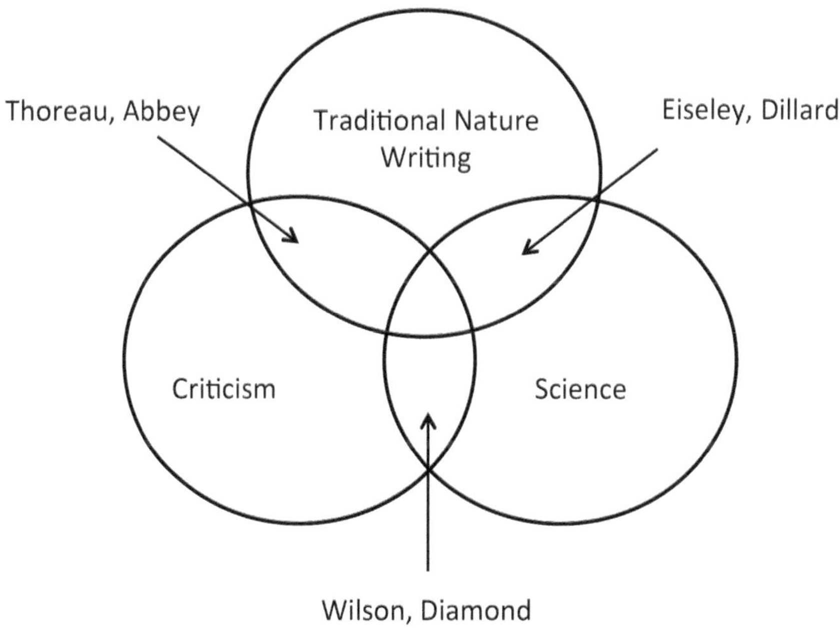

Figure 7.1. Elements of Environmental Literature, copyright 2014 Paul Lindholdt.

to Ralph Waldo Emerson, and forward to the nature-advocates of the present day. But plenty of novelists (e.g., Edward Abbey, Wendell Berry, Leslie Marmon Silko, and Ursula LeGuin) and poets (e.g., Mary Oliver, Theodore Roethke, Pattiann Rogers, and Gary Snyder) also have generated enduring environmental belles lettres. Yet it is within non-fiction nature writing and environmental literature, both primary and secondary works, that our definitions need to accommodate the literature of the culture clash, the literature of the WUM.

BACKLASH IN THE NORTH CASCADES

Late in 1993, Mitch Friedman and I published *Cascadia Wild: Protecting an International Ecosystem*. That book, a collection of essays we wrote along with other activists and scientists, studied the ecology of one of North America's largest contiguous wilderness areas and proposed that a Cascades International Park be established and managed jointly by the United States and Canada. This proposal came about after a coalition of environmental groups on both sides of the border discovered that the piecemeal approach to conservation in the North Cascades was ineffective. Canadian Crown and U.S. Federal lands butt against private inholdings on both sides of the border and create a bureaucratic tangle that fragments wildlife habitats. Despite this fragmentation, there is geographical integrity enough in the North Cascades to warrant a park plan. The governments of both countries wisely have set aside terrain for protection over recent decades. On the Washington side are the North Cascades National Park, three recreation areas, and seven wilderness areas. Other lands are semi-protected through federal forest plans. In British Columbia there are Manning Provincial Park, Cathedral Provincial Park, and two recreation areas. All of these names on a map might give the appearance of discontinuity, but the area is a remarkably intact and contiguous ecosystem—a diverse system of plants and animals thriving in adjoining indigenous environments. Is *Cascadia Wild* nature writing? Some readers believed it is, despite its reliance on disparate research approaches and methodologies and its often-polemical approaches and tones.

My partner in *Cascadia Wild*, Mitch Friedman, is the founder of Conservation Northwest, formerly the Northwest Ecosystem Alliance, before that the Greater Ecosystem Alliance, of Bellingham, Washington. The book was his brainchild. Washington State organizer for Earth First! when we met in 1988, Mitch co-edited and wrote key chapters. I contributed a single chapter on the impacts of livestock grazing on public rangelands that concluded with a case study of cattle allotments in the Okanogan National Forest. Because my interest lies chiefly in language and literature, I approached my chapter by deconstructing government reports on livestock grazing, a deconstruction reproduced here in the chapter "Literary Activism and the Bioreogional Agenda." We solicited studies from scientists, activists, academics, independents, and conservationists. Our guiding principle in the book is conservation biology, a science designed to retain and restore ecosystems that industry, technology, and bureaucracy have sundered.

After the book came out, it began to get some favorable press in parks and conservation magazines, but backlash by regional WUM organizers was bitter and swift. The first two critics of the park proposal, Washington State citizens Don Kehoe and Homer Bakker, toured the Northwest with a fugitive Christian Patriot group called Citizens for Environmental Justice. That group distributed literature from the Militia of Montana and promoted a theory that the United Nations was poised to take over the North Cascades of Washington. Our proposed park, they claimed, was intended as part of a global plot to fragment the United States and install a one-world government. If this international scheme were ever to succeed, they warned, internationals would control one-quarter of Washington State and kick out current residents. Beware, Kehoe cautioned—the borders of such a compound utilize electronic fortifications patrolled by the CIA (Pryne A1). Imagine our surprise, we writers and editors, scientists and culture critics, to be named pawns of a one-world government. Kehoe asserted that owners of private lands in the North Cascades, after arriving at their remote cabins, found themselves accosted by Forest Service rangers who claimed to have satellite-tracked their arrivals. Kehoe forgot to get the landowners' names, but he later carried his message of espionage and fear to Maltby, Washington, for a recruitment meeting for the Militia of Montana (deArmond).

Overt anxiety about government intervention into private lives, a cornerstone of right-wing ideology, looms especially large in the rural West where natural resources have underpinned certain slumping economies. In those places WUM organizers, often funded by industries that have much to gain by relaxing environmental regulations, make greatest progress where people are the least educated (Deal). Connections are tight, according to my research, between gay rights opponents, militias, conservative think tanks, and the leaders of the powerful WUM who exploit irrational fears to further their economic and political agendas (Ramos 103-06; Lindholdt, "Wise" 10–13). While some of these examples might seem dated, the movement has not simply disappeared; it has changed shape, gone mainstream, and has been entering the halls of Congress.

The rise of private American militias, whose theories of conspiracy were so well documented in the wake of the Oklahoma City bombing, is attributable in part to fear and hatred of outsiders and outside laws. American Jewish Committee hate-group specialist Kenneth Stern, in his book on the militia movement, devoted a chapter to the militia backlash against environmental regulations. In Stern's analysis, private militias in the mid-1990s "attracted people frustrated over environmental regulations, especially in the West" (119). The trend toward county supremacy was a consequence. County officials and private landowners claim federal lands are actually theirs by custom and culture and therefore subject to their sovereign jurisdiction. They defend this viewpoint by using the legal argument of adverse possession, which says that a landowner who lets his land lie idle for a given amount of time effectively surrenders claim to it. According to the custom-and-culture argument, which stems in turn from arguments of adverse possession, individuals who have grown accustomed historically to using public lands are entitled to lay claim to those lands, if those same lands are now essential to upholding their culture and if the government has let the land lie idle. To deny them rights to graze their cattle on those lands, effectively, is to deprive them of their customary civilization. Counties in Idaho, Nevada, and New Mexico have gotten publicity for decades by doing their best to seize federal lands. The Posse Comitatus, a Christian militia group whose name means "power of the county," shares such tactics. Posse Comitatus members formed common cause in the 1980s with white supremacist organizations like The Order, whose members are dead or in prison after murdering Denver talk show host Alan Berg in 1984 (Corcoran 42). At least seventy U.S. counties have passed county-supremacy laws that challenge federal jurisdiction on federal lands. Cliven Bundy refused to pay grazing fees for twenty years and became a darling of right-wing media early in 2014, until he began spouting racialist beliefs, whereupon most of his supporters abandoned him to his own devices.

The nativist beliefs of some WUM members link them to white supremacists. At a 1994 meeting in Whatcom County near Bellingham, Washington, citizens assembled to explore the formation of sovereign "Freedom County." Organizer John Stokes addressed the threats of the Growth Management Act passed by Washington voters in 1993 to curb urban sprawl and conserve wetlands. Stokes warned that "job allocation limits" will forbid new employment in certain areas, but wealthy counties will be able to buy out of this mandate with what he called "mitigation fees." Poorer rural counties, he reasoned, unless they claimed to become self-governing, ultimately will be "dumping grounds" for low-income folks and "Haitian, Chinese, and Russian immigrants, uneducated people, people with HIV, the criminally insane, sexual predators," and others (deArmond). On the east side of the Washington Cascades, especially in Ferry and Stevens counties, WUM militants have spread unrest about an impending "new world order." Kettle Falls resident Glenn Rowe opened a meeting in February 1995 in Colville by denouncing "commie Nazis" who had infiltrated the federal government to take over the United States— "a Christian nation." A second speaker, Fred Kisman, asserted that socialist journalists, by discussing counties as "regions" instead of as autonomous political entities, are taking steps toward abolishing traditional borders. "We're being psychologically brainwashed," Kisman insisted. To prove it, he pinned up an ecosystem map entitled "Nature Has No Borders," from a conference being hosted by the University of Washington (Marie A1).

That conference, cosponsored by Greater Ecosystem Alliance in 1994, returns us to the zealous backlash to the park proposed in *Cascadia Wild*. Don Kehoe, the Christian patriot speaker who claimed the Forest Service had been tracking private landowners in the North Cascades by satellite, spoke again on January 16, 1996, in Bellingham, Washington, to a group called Citizens for Liberty (deArmond). A former landscaper, Kehoe served on the board of the now-defunct Snohomish County Property Rights Alliance, the same county where the massive Oso landslide took the lives of forty-two people in March 2014, probably due to irresponsible logging uphill. The topic of Kehoe's speech at the Citizens for Liberty meeting was "The CIA, Toll Roads, and Mind Control." He distributed papers from various public agencies that allegedly describe a plan to implement an Intelligent Vehicle Highway System (or IVHS). This supposed technology entails a network of highway sensors and "smart" cars. Such fears of government incursions have gotten the ears of legislators, which might bode badly for the autonomous car that Google hopes to bring to market. As research on the Oso landslide may eventually demonstrate, WUM activism can palpably impact human lives.

Kehoe wove his discoveries into a conspiracy theory involving the CIA, the United Nations, and the new world order. According to Paul deArmond, in his audience, Kehoe claimed that IVHS technology should allow vehicles to be remotely tracked, monitored, and controlled by satellite transmissions. Ostensibly the microchips to bring about this control are already in place in late-model cars. Those who followed Oklahoma City bombing reporting will remember suspect Timothy McVeigh's claim the U.S. military had surgically implanted a microchip tracking device in his buttocks. One of the links in Kehoe's sinister chain of technology is his discovery of the Cascadia Project, which he claims is participating in IVHS discussions. "Cascadia," which by now has devolved into marketing jargon in the Northwest, denotes everything from a bioregion to a soft drink. The Cascadian Independence Project, however—in fact a planning effort spanning the I-5 corridor from Portland to Vancouver, British Columbia—seeks to accommodate present and projected trends in commerce, transportation, and urban growth.

These extremist mouthpieces of the WUM represent the movement no more than monkey-wrenching and sabotage characterize environmentalists. It is instead in the more moderate and reflective counter-environmentalism that the simultaneous threats and opportunities reside. I characterize them as threats because they impede the valuable and necessary work of conserving species and ecosystems, because they confuse students with counter-scientific claims, and because ignoring them imperils perceptions of fair play. I characterize them as opportunities in their potential to expand the abilities of environmental students, writers, and activists to confront their rivals and clarify thought.

ORIGINATING A MOVEMENT

The term "wise use" originated with Gifford Pinchot, U.S. Forest Service Chief under Theodore Roosevelt. Scarcely a friend of the environment, Pinchot supported a 1913 dam in the Hetch Hetchy Valley near Yosemite, a valley that now is stoppered like a bathtub to make the city of San Francisco bloom. Pinchot argued that the project was an improvement upon nature, but the dam in fact proved to be chiefly about political power. John Muir opposed the project, which sparked the founding of the Sierra Club. An earlier boondoggle that empowered the proto-movement toward the WUM was the 1864 public land giveaway under Abraham Lincoln known as the Northern Pacific Land Grant, which is the subject of a book by Derrick Jensen and George Draffan titled *Railroads and Clearcuts*. Such historical precedents remind Americans that government long has combined with industry to push through environmentally destructive projects, although we might be tempted to think of the WUM as a trend that is recent and corporate only.

To see the WUM as just another skirmish in what have come to be called the culture wars dividing America in recent decades is dangerous and wrong. The problems have been around a lot longer than that. In writing his play *An Enemy of the People* in 1882, Henrik Ibsen grappled with some of the same scientific and economic rifts that currently divide environmentalists and the WUM brigades. Insofar as human cultures construct nature, Ibsen's can be considered nature writing in its broadest sense. Brothers Thomas and Peter Stockmann are clashing over the economic fate of a small Norwegian town. Peter is the wealthy magistrate and mayor, while his brother Thomas is a physician, a man of science. Together they are building a health institute from mineral springs that are reputed to have curative powers. The mineral springs are drawing sick people from far away—a ninteenth-century business trend that T. Coraghessan Boyle lampooned in *Road to Wellville*. Can Boyle's novel justifiably be called nature writing? It criticizes cultural trends that impact nature; it is satirical, not rhapsodic. Can nature writing rightly satirize? Carl Hiassen's send-ups of boosters and developers in Florida show us how. The same is true of climate science today.

Just as the WUM issues denials and uses counter-science today, so the leaders of the fictional Norwegian town deny the science underlying the findings of physician Thomas, who cries with exasperated irony, "Who am I, a mere scientist, to tell politicians where to build a health institute!" (36). Unlike the rest of the community, he has the training to discern the truth through the lens of his microscope. Exactly like the other townsfolk, though, the matter makes him passionate. To Peter's economic arguments about the "source" of the town's "most important industry," Thomas responds, "That source is poisoned, man. We are getting fat by peddling filth and corruption to innocent people!" (58). Peter follows through on threats to declare his scientist-brother an enemy of the people. The people then pelt Thomas's house with stones and assault his children. Peter's rhetoric makes over nature in a business paradigm. He decries Thomas's scientific report as "the mad dream of a man who is trying to blow up our way of life! It has nothing to do with reform or science or anything but pure and simple destruction!" (74–75). More than a century later, many resource-dependent communities in the rural West are echoing Peter Stockmann's cry, believing they are victims of misinformation spread by scientists and activists whose agendas are not objective but politically driven.

Aldo Leopold, author of the classic *Sand County Almanac,* remarked upon a trend that likewise characterizes our own wise-use scuffles. Leopold made his observations in a chapter entitled "The Land Ethic," which was delivered first as a paper before the American Association for the Advancement of Science in 1933. "When the private landowner is asked to perform some act for the good of the community," Leopold wrote, that landowner commonly assents "only with outstretched palm" (202). Leopold lamented that only "economic self-interest" provided impetus to the farmers of his place and time to conserve natural resources (203). Unless they can be tempted by promises of profit or compensation, Leopold concluded, most individuals and corporations simply will not agree to abide by environmental codes and regulations. The truth of Leopold's assertion is evident today in the pollution credits and tax incentives our government uses to gain corporate compliance with environmental laws. Leopold's paper, originally delivered as a public speech, is nature writing, if we willingly expand our definitions.

Aldo Leopold lamented that economic imperatives hamstrung conservation efforts in the 1930s. So did David Brower in the 1971 book *Encounters with the Archdruid* by John McPhee. McPhee expertly chronicled the encounters between conservationist Brower and three counter-environmentalists, three WUM precursors. Brower debates the merits of mining with Charles Park, a mineral engineer who believes our economic wellbeing relies on extracting metals from the planet wherever they be found, centering on the North Cascades Park of Washington State, the locus of the international park that Mitch Friedman and I proposed. There the men hike together. On Cumberland Island in South Carolina, Brower encounters real-estate developer Charles Fraser, and on the Colorado River he meets and debates Floyd Dominy who impounds big rivers for a living. The book presages the terms of the WUM in the U.S. and the Share movement in Canada. All three men insist that they are practicing the wisest use of natural resources.

Developer Charles Fraser echoes the Gifford Pinchot phrase when he asserts, "The government has a perfect right to condemn my land here if it thinks its use is wiser than mine" (McPhee 140). WUM leader Ron Arnold of Washington State co-opted the Pinchot phrase in 1988 and adapted it for use by the counter-environmental forces (Lewis 16). Charles Fraser, twenty years earlier, practiced the same co-optation as WUM founder Ron Arnold—a practice known widely today as greenwashing, a common modus operandi of public relations firms—by naming his new development "the Cumberland Island Conservation Association" (McPhee 138). In studying tactics of the counter-environmental organizations, Stanford biologists Paul and Anne Ehrlich dub such practices "aggressive mimicry" (16)—borrowing from biology a term to name the co-optative tactics whereby the parasite or predator adopts the colors or behaviors of its host or prey. Such cultural clashes are battles of words, of ideology, and of science. Developer Fraser decries environmentalists for the way they "worship trees and sacrifice human beings to those trees," resembling pagan druids of old (95). McPhee's book is a best-selling classic of nature writing, even though most of its approach is journalistic.

NAMING THE PLAYERS

Conventional wisdom on the WUM has viewed it as a collection of industry front groups. In the mid-1980s, corporations that profited from extracting natural resources hired organizers to go into rural Western communities and stir up distrust of regulations. Honda and Suzuki proved to be financial mainstays of the Blue Ribbon Coalition, for instance, a WUM group founded in Idaho that agitates to open wilderness and park lands to snowmobile and motorcycle riders (Deal 33–34). That same off-road vehicle industry also keeps squads of lobbyists on salary to badger beltway politicians into voting against environmental laws. Blue Ribbon was one of the original participants in a June 1988 conference that was held in Las Vegas to ally disparate WUM interests. From that conference emerged "The Top Twenty-Five Goals of the Wise Use Agenda," including watering down the Endangered Species Act, defunding the National Parks by allowing their privatization through companies like the Disney Corporation, and a host of other such reformations of nature. The Las Vegas conference has now become an annual event. And the spate of backlash books appearing since that conference is itself an emerging form of nature writing, intended for a conservative readership mindful of the culture clashes that pivot on natural resources. Ehrlich and Ehrlich dub such books part of the "brownlash," a nonce word that suggests that the backlash against the environmental movement will turn the planet brown, and they agree that teachers of American nature writing imperil their own interests if they ignore such books of counter-science (217).

A second segment in the wise-use coalitions are conservatives and libertarians who respond contemptuously to government intervention into the economy, politics, and religion. Perhaps the most visible and active among such conservatives has been the Rev. Sun Myung Moon (1920–2012) and his Unification Church. The extant empire of the Rev. Moon is immense, marketing everything from colas to guns. Its chief front group, the American Freedom Coalition, works to open the Alaska National Wildlife Refuge to oil exploration (Helvarg, *War* 260). Very similar in his wealth and conservatism and distrust of environmental regulations is Lyndon LaRouche, convicted tax felon and erstwhile Nazi Party leader whose followers spearhead campaigns to discredit ozone-depletion theories and other forms of science that block his platforms and his profits (Helvarg, "War" 254-58). Buell traces the renaissance of the WUM to politics as well. "Sadly deprived of this old target with the fall of communism and the end of the 'evil empire,' the right chose environmentalism to fill the gap" (22). I call it a "renaissance" because, as I have been trying to suggest, the movement has been around a long time.

A third sector of support for wise use includes grassroots activists. Here the mix of ideologies and priorities gets very tricky to depict. While industry organizers recruit some grassroots activists, others are as independent as the wind. Here common cause is made with extreme Christian fundamentalists, conspiracy theorists, tax resisters, militias, survivalists, and even white supremacists. The movement, as the late researcher Paul deArmond observed, is less a movement than a hydra, that monster of Greek mythology which astounded with its capacity to grow two heads to replace each one cut off in battle. Just as that mythical beast regenerated itself, so the extremist fringes of the WUM and the militant right display the ability to resurface with redoubled energy after each defeat. Alan Gottlieb, who cofounded the WUM with Ron Arnold, is "a convicted felon who once lost the ability to possess guns" (Sugarman). Gottlieb is President of two non-profit corporations that constitute the most potent U.S. pro-gun forces besides the NRA: the Second Amendment Foundation and the Citizens' Committee For the Right to Keep and Bear Arms. While Gottlieb is no grassroots organizer, his advocacy for guns has made him a guru for many of those who are.

WUM coalitions unite in the goal of repealing or watering down environmental regulations. The most prominent leaders in my corner of the Pacific Northwest are Arnold, Gottlieb, and Chuck Cushman. All of them live in Washington State. Despite their dedication, Washington remains a very blue state whose citizens voted down a stringent counter-environmental property-rights measure. Washington Referendum 48 would have made governments (i.e., the taxpayers) compensate landowners if zoning or environmental laws should diminish property values. Washington's two women senators and governor all are Democrats; the latter vows to wean the state's utilities off coal. Ron Arnold—the former public relations consultant to industry who gave the movement its name when he appropriated the 1907 Gifford Pinchot definition of conservation as "the wise use of resources"—openly aspires to "'systematically destroy the environmental movement,' which he believes to be 'polluted with a hatred of humans'" (Lewis 15–16). Chuck Cushman founded and still runs the American Land Rights Association and has grown well known for his fax alerts that urge his followers to lobby lawmakers to vote against incursions by the Federal government and The Nature Conservancy, groups that often, he says, work in cahoots. The latter "would buy from financially distressed landowners, then turn the land over to the Federal government. Time after time this happened, quietly, secretly and silently they helped undercut the community" (U.S. Senate). Nor is the backlash as I define it here confined to the American West or the United States. Royal Dutch Shell in Nigeria shared enough complicity in the hanging of environmental activist Ken Saro-Wiwa in 1995 that it settled out of court with his family in 2009 for $15.5 million. Rowell dedicated his book *Green Backlash* to Saro-Wiwa.

The Sahara Club gave us a taste of the sort of hate speech that the WUM can degenerate into. The Sahara Club began as a band of angry California dirt-bikers deprived of racing across a certain desert where tortoises were found crushed (Deal 87–88). The club's chief function was to breathe fear into the conservation movement. In 1994 it issued an "Agenda to Attack Environmentalists." "Consider this your guide book," the document begins, "to effectively battling New Age nuts, militant vegetarians, anti-gun pukes, animal rights goofballs, tree worshipers, new world order pushers, human haters, pro-socialists, doomsdayers, homosexual rights activists, militant feminists and land closure fascists!" ("Sahara Club's"). The guidebook offered dirty tricks to intimidate government officials and environmentalists, including smashing headlights at environmental gatherings with ball bearings wrapped in a sock—a weapon that many readers would have recognized as a de facto blackjack. Organizing became more dangerous for activists and bureaucrats, especially those who perform fieldwork. A former student of mine held a controversial job as rural-outreach coordinator for the Idaho Conservation League that carried him into hostile hotspots in North Idaho to speak to the public and to give slide shows. He did not last long, partly out of concern for his safety and his wife and children. It did not avail him that his name is Guadalupe Flores.

Two examples of counter-environmental intimidation should suffice. At a public hearing in Everett, Washington, Audubon Society activist Ellen Gray testified in favor of protection requirements for wildlife and wetlands. Two men approached her afterward. According to Gray, one placed a hangman's noose on a nearby chair, saying, "This is a message for you." He also distributed cards with a picture of a hangman's noose that said, "Treason = Death" on one side, and "Eco fascists go home" on the other. The other man told Gray, "We have a militia of 10,000. If we can't get you at the ballot box, we'll get you with a bullet" (Helvarg, "Anti-Enviro" 724). Similarly, in the fall of 1994 Oregon activists Ric Bailey and Andy Kerr were burned in effigy in rural Joseph, Oregon. Although these examples are by now some twenty years behind us, organizers in the hinterlands tell plenty of tales of such misbehavior still today. Besides acts of random (and, some would say, mostly toothless) intimidation, the widespread national tactic of WUM groups is to kindle discontent regarding public and private land-use policies and to work with think tanks. Congressman Richard Pombo, a six-term Republican legislator for California, emerged as a WUM stalwart and vocal opponent of the Endangered Species Act. He published a book riddled with fallacies and titled *This Land is Our Land*. But Pombo is only one of dozens of writers who have turned to literature and scholarship as useful tools to combat what they see as the excesses of mainstream environmentalism.

WRITING THE BACKLASH

Certainly the most active and provocative writer in the WUM is Ron Arnold, a former Sierra Club member. He has written eight books that assail environmentalism and environmentalists, edited ten others, and published some three hundred magazine articles. Arnold embarked on his literary career with a biography of James Watt, Interior Secretary under Ronald Reagan. In that 1981 book, he not only leveled ad hominem attacks but also applied some ambitious semiotic analyses to the concept of "wilderness." He saw the social construction of wilderness as comprised of "images and emotions of profound connection to the earth, of pristine primal peace, of nature in perfect harmony, of a time and a place unmarred by the meddling hand of man" (xiv). By any definition, Arnold practices environmental writing. He seldom recurs to science, and he scorns rhapsody, but he builds a culture criticism in his own terms through his antipathy to the movement. Nature is for him a confused site of dangerously skewed perceptions. With equal virulence he disdains journalism, about which he writes, "Drama, the active pursuit of conflict, has become a basic convention of newswriting, and that is not likely to change. Every page of the newspaper is the drama page. Every newscast is a soap opera" (210). Arnold subsumed the life of James Watt to a critique of cultural issues surrounding him.

A frequent refrain of WUM writers generally, and of Arnold in particular, is that economic ruin ensues from environmentalism. Hence the title of a book he co-authored in 1993, *Trashing the Economy: How Runaway*

Environmentalism is Wrecking America. (His title alludes to 1990 literary milestone of the WUM titled *Trashing the Planet*, by the late Washington State governor Dixie Lee Ray, an advocate of nuclear power and a chair of the U.S. Atomic Energy Commission.) Arnold's co-author of that book and cofounder of the movement, Alan Gottlieb, earned a degree in nuclear engineering, has written nineteen books that mostly advocate gun rights, and additionally presided over the Center for the Defense of Free Enterprise until he handed the reins to Arnold. *Trashing the Economy* was self-published, possibly suggesting that even supportive presses found it unsound.

Disputes over nature and natural resources in America are battles in an ongoing culture war, and the WUM writing hastens the pace. The disputes pit environmentalists, scientists, and bureaucrats against what Anne and Paul Ehrlich termed as "representatives of extractive and polluting industries who are motivated by corporate as well as private property rights activists and right-wing ideologues" (15). In his book, erstwhile legislator Richard Pombo centered the dispute within an "eco-federal coalition" that has given rise to "a growing body of federal, state, and local regulations and . . . open-ended Supreme Court decisions without standards or constitutional foundation" (7). The tilt of the Supreme Court toward the conservative side of the aisle must have mooted his worries some. Paul Ehrlich, a Stanford University biology professor and leading scholar in the environmental sciences today, is best known for his 1969 book *The Population Bomb*, which has been roundly ridiculed by WUM authors for the cataclysmic scenarios it painted of food shortages resulting from population pressures. Pombo, who wrote his book in the first-person singular, names his nemeses "environmental zealots who believe that mankind brings only harm to his environment" (5). One of my interests in examining such extremes is traceable to the niche theory. Deriving from biological ecology, the concept of the niche as it manifests in sociology represents positions or functions of groups in larger community environments. The niche theory poses key questions about the survival of groups under changing environmental conditions and how they interact with competition. Will the WUM survive in its present form for another generation?

Both sides of the culture war over environmental issues accuse each other of the same politicization of science, and that divide is only deepening with the mainstreaming of debates over global warming and climate change. The central premise of the Ehrlichs is that contrarian or non-consensual science, a body of claims about the natural world that rejects "the deepening consensus among scientists" (2), is being used "to bolster a predetermined worldview and to support a political agenda" (12). Anyone who follows congressional hearings and the scientific consensus about climate change will be familiar with such agendas. Contrarian science, strengthened by an "overall lack of scientific knowledge among

United States citizens" (13), poses a grave threat insofar as it is "being adopted and used as a basis for critical policy decisions" (201). In short, politicians like Sen. James Inhofe on global climate change today often found their policies on non-consensual scientific conclusions that arise from right-wing ideologies. One result is turbulence in politics. "Experts are increasingly drawing connections between climate change and the erosion of the social contract between citizen and government, between the people and the land" (NPR). Complementing the 848 notes in the Ehrlichs' book are two appendices, the first of which analyzes "Brownlash Literature" of the sort produced by Gregg Easterbrook and Stephen Budiansky, whose chief interests have been to debunk green alarmists and to bring a new optimism to the environmental outlook. The Ehrlichs' second appendix, "The Scientific Consensus," names and numbers international scientists who agree in trepidation about the dismal state of environments today. Since the Ehrlichs published their book, the rift between the WUM and environmentalists has grown deeper with concerns about catastrophic weather patterns and the entry of legally liable insurance companies into public dialogues. Anne and Paul Ehrlich write well and demonstrate a gift for analogies that speak to lay readers. Their ambitious book was a call to activism, a plea for scientists and citizens to "'tithe to society'—spend 10 percent of their time trying to make the human endeavor more sustainable" (205). Those engaged in organizing, in my region against the increase of coal- and oil-tanker trains, for instance, distinguish between themselves and WUM players who they say have palpable economic interests at stake.

The former Congressman Richard Pombo wrote a more private response to the ongoing changes in American public policy. His book, coauthored with journalist Joseph Farah, culminated by promoting Pombo's reform measures for the Endangered Species Act, or ESA, signed into law by President Richard Nixon more than forty years ago (188-93). Pombo recounts his family travails when a railroad right-of-way through their ranch in California violated their perceived property rights (114-115). Federal thugs, Pombo charged, now oversee the 1973 ESA by wielding "their own private, armed, enforcement agents" (27). Essentially an anti-ESA tract, *This Land Is Our Land* borrowed from a proliferating WUM literature as well as from studies by the Cato Institute, opinions by Supreme Court justices Clarence Thomas and Antonin Scalia, and from the anonymously authored article "Eco-Facism" in a paper tellingly entitled *The Freeman* (100). Other aspects of Congressman Pombo's book included reasoning fallacies, slipshod editing, and lack of documentation. Lake Tahoe is near "Loss Angeles" (100), while "today's miner's are very land-efficient" (108). Congressmen Murkowski, Stevens, and Young are said to hail from Arkansas, not Alaska (159). Eugene Hussey—a WUM poster figure who claims to have been harassed by federal agents when he "found" a dead wolf on his remote Idaho ranch—was said to have been both seventy-four and seventy-five years of age at the time (27, 170). In a flagrant appeal to ignorance, Farah and Pombo dismiss rising extinction rates by noting, "some of these species may carry organisms or bacteria that could be extremely harmful to humans or other life. Maybe their extinction will save us!" (54). A false analogy arises when they ask, Sagebrush Rebel-fashion, "Are the people who give us slow mail service, low SAT scores, and the 1040 tax form best suited to control millions of acres of land?" (182). Nowhere did the authors address the role of science in shaping public policy. Yet the center of WUM disputes is the politicization of science that grounds so much public policy. There the key issues reside. There the opportunities for ecocritics abound.

THE INCIDENCE OF DDT

Ibsen, Leopold, and McPhee wrote classics of nature writing and its sibling-genre, environmental literature, by critically examining the interface of environment and free enterprise. In doing so they highlighted some of the culture clashes that would occasion the later rise of the WUM. But none of those books had the influence of Rachel Carson's *Silent Spring* (1962), which singlehandedly caused a ban on DDT in the U.S. (DDT still is manufactured in the United States and brokered to developing nations). What Upton Sinclair's *The Jungle* (1906) did for meat safety in the United States, Carson did for American nature writing and the environmental movement. Accordingly the backlash against Carson gained a strident life. American chemical companies worked hard to malign her. That malignity, coupled with specious counter-science, can be traced back to interviews conducted and research published by the late Dr. Dixie Lee Ray—a former governor of Washington State, first woman to chair the Atomic Energy Commission, and professor of marine biology.

If Ray had a high profile as a regional leader in her time, she is more significant today for the way she lent her credibility to counter-scientific ideologies when she became a spokesperson for the WUM. Pundits like Rush Limbaugh used her work as ammunition for their own counter-scientific claims. Those of us baby-boomers raised in Washington State can recall Dr. Ray's staunch advocacy of nuclear energy and her ability to baffle common sense and scientific evidence when they contradicted her beliefs on natural-resource conservation. In her books and interviews, Ray maintained that the preservation of wetlands is a foolish error perpetrated at the behest of environmentalists. Those environmentalists privilege dangerous pest and predator species above humans.

Recriminations against conservationists have become the stock-in-trade of the WUM. What set Dr. Ray's arguments apart were the spuriousness of her reasoning and the intricacy of her errors. Wetlands protect mosquitoes, she said. Because scientists have supported public policies that favor the protection of wetlands, mosquito populations have boomed, and the result is unheralded and unsustainable increases in insect-borne diseases like encephalitis. Even worse, Dr. Ray claimed, "'the central valleys of California are areas of endemic malaria'" (Calahan 11). This assertion, made in 1993, reiterated the 1990 claims in her book *Trashing the Planet*, which added yellow fever to the list of hazards exacerbated by the ban on DDT and protection of U.S. wetlands. Yet biologists, scientific research, and the bureaucrats contradict her. "Ray's alarmist claim that environmentalists are responsible for malaria epidemics is both contrived and grotesquely dishonest" (Callahan 15). The astounding feature of this hoax, perpetrated on a scientifically gullible public by an educated public servant speaking from her bully pulpit, is that it took so long to be uncovered. Nor does the hysteria end there. The wetlands-malaria fraud continued to find sympathetic audiences at WUM conferences and credulous readers in right-wing periodicals like the conspiratorial *Executive Intelligence Review* and *21st Century Science and Technology*, both periodicals projects of Lyndon LaRouche. One writer in the latter publication argued that *Silent Spring* was evidence of a plot to discredit pesticides and chemical producers, to elevate disease levels, ultimately to hasten death rates as part of a Malthusian scheme to roll back global human populations (Edwards 25). Thomas Malthus, ecocritics will certainly remember, was the 19th-century English economist who prophesied that human populations would continue increasing at rates faster than available subsistence, thus causing poverty and degradation if not checked by moral restraint, disease, famine, war, or other natural or human-induced disasters. His name, almost always rendered adjectivally, has become synonymous, in WUM locutions, for irrational alarmism.

CONCLUSIONS

The backlash to the Cascades International Park we proposed in *Cascadia Wild* is evidence of a larger movement to discredit ecologists and environmentalists. Fueled by reactionary policies, funded by corporations that rely on access to natural resources to plump their bottom lines, the movement successfully spurred and still galvanizes people in rural communities. Such people target the environmental movement mistakenly, for the changes those rural communities are resisting will prove inevitable. As populations burgeon, in the western states especially, economies in rural communities are undergoing inevitable transitions. In this process extractive industries like logging, mining, and ranching are coming to be supplanted by recreation, tourism, and retirement homes. Such industries have relied traditionally upon federal subsidies and land giveaways to keep them solvent, which accounts for the demonization of governments and environmentalists by the WUM now. Newcomers to the West—taxpaying citizens and voters—statistically favor environmental safeguards as well as cutbacks in tax-supported public welfare.

As western economies undergo vast economic changes, the cultural constructions of nature change with them. Seen through the lenses of the WUM, nature is less a source of spiritual solace and emotional refuge than of material value and constitutional liberty. According to such formulations, nature comes again to recall the site of God's injunctions to Adam and Eve to "Be fruitful and multiply, and replenish the earth and subdue it: and have dominion over the fish of the sea, and over every living thing that moveth upon the earth" (Gen. 1:28). That is to say, nature once more becomes a place where humankind may and must exert its will to extract material goods and prosper from its extractions.

Seen this way, the WUM is more than a backlash against changing economic trends, more than antithesis to the bureaucrats perceived to be its servants. The wise-use view alters perceptions of nature by constructing those perceptions anew. It is fitting and proper that current definitions of the nature writing and the canon should shift to consider the counter-scientific ideologies of the WUM. It is also fitting that ecocritics may and should expand our interests beyond the rhapsodic. By reading Ibsen, Leopold, Carson, and McPhee (if not Arnold, Pombo, and Ray) as texts that address an inchoate counter-environmental movement, we can expand our definitions of environmental literature and appreciate the range of ways that nature continues to be socially constructed today.

NOTES

1. Evernden's analyses are varied and subtle, but they are perhaps best encapsulated in his observations regarding "the camouflaging tendency of the acculturated consciousness, which renders direct encounter [with nature] impossible most of the time" (122). Claire Lawrence concurs, and goes even further, when she notes, "The designation of the natural and the reading of the natural is always cultural" (165). The materialist turn in ecocriticism offers welcome counterpoint to such social-construction paradigms.

2. Barry Lopez, for instance, concurs: "writing that takes into account the impact nature and place have on culture is one of the oldest—and perhaps most singular—threads in American writing" (1). For poetic expression of this notion, see "The Determinations of the Scene" (32–33) and "The Significance of Location" (38) by Pattiann Rogers, in the former of which the speaker traces children of the desert, forest, and seaside, and for whom place constitutes "The underlying dominion unacknowledged / In [one's] approach to the cosmos" (38).

EIGHT

Greening the Dramatic Canon

Henrik Ibsen's An Enemy of the People

Works of environmentally inflected literature in every genre except drama abound. Notably in the literatures of America whose tracts of virgin lands have charged the imaginations of writers since colonial days, readers find an embarrassment of riches, ecstatic essays and poems especially, which rely on natural scenes. But *plays* that treat nature and environment, in all ages and nations, are rare. One can account for such generic disparity by remembering that drama lends itself to neither the rhapsody nor the scientific detachment that compounds environmental literature. Moreover, ecological crises are far more common now than before the Industrial Revolution, before the great world plays were written. If today's playwrights fret that ecological dramas might be judged as propaganda, they need only recognize the way a spate of Hollywood films, based on true stories, triumph by translating ecological themes to the silver screen. All of these films take off from true events. *Silkwood* concerns a worker at an Oklahoma plutonium plant who died in a suspicious car accident before she could blow the whistle about hazardous conditions at the plant. *A Civil Action* tracks a slick Massachusetts lawyer who lost all his status and cash when he chased justice in an industrial pollution case. And *Erin Brockovich* valorizes a legal clerk whose persistence convicted a guilty Pacific Gas & Electric utility and won a hefty settlement for her sick clients and her firm. While our technologies and populations burgeon, patrons of movie theaters are paying to appreciate environmental problems transmuted into cinematic art. At the same time, more secondary teachers are looking at environmentally inflected arts and sciences in order to make their curricula more interdisciplinary, coherent, and relevant to student needs.[1]

The films whose plots I sketch above bear close comparison to *An Enemy of the People,* Henrik Ibsen's powerful but unheralded 1882 play whose green themes pertain as much to our time as they did to Ibsen's. The long production history includes one 1898 run in Tokyo where, according to Emeljanow, the play "formed part of a protest against chemical spillage—the play's first environmental application" (101). Here is a plot summary whose details should carry contemporary familiarity. The brothers Thomas and Peter Stockmann clash over the economic future of a Norwegian town. Peter is the wealthy magistrate and mayor; Thomas is a physician, a man of science. Together they are building a health institute from mineral springs that are reputed to have curative powers. The mineral springs already are drawing sick patrons from far away. Promoter Peter is booming with enthusiasm for the enterprise. The last thing he wants to hear from his scrupulous brother is that an upstream tannery has polluted the springs. But in fact diseases caused by tanning animal hides have proliferated in the stream. When Thomas proves that microscopic bacteria leaching from the tannery are infecting patients—when he argues that costly repairs will be necessary before sick bathers can patronize the pools—Peter rallies the townspeople to discredit his brother's threatening message. The people of this small town, their reputations and newfound prosperity in jeopardy, turn furious. They silence Thomas, pelt his house with stones, assail his children, and proclaim him an enemy. Homeless with his family at the end, Thomas Stockmann stands unbowed.

American playwright Arthur Miller adapted this play for contemporary audiences in 1950,[2] and he acknowledged its growing environmental relevance in 1989. His adaptation of the play, with all its strengths and quirks, its high spirits and idiomatic obsolescence, has spawned a body of commentary on its own. That adaptation, from what Miller termed in the Preface "a word-for-word rendering of the original, done by Mr. Lars Nordenson" (11), was published amid hearings by the House Un-American Activities Committee (recognized officially from 1934 to 1977). Wisconsin Sen. Joseph McCarthy led those activities. "In a 1950 speech," Stanley Schulz reminds us, "McCarthy entered the public spotlight by claiming that communists had 'infested' the State Department, dramatically waving a sheet of paper which purportedly contained the traitors' names."

Playwright Arthur Miller, who had himself gained a reputation as a socialist and was required to testify before the Committee in 1956, protested artistically against McCarthy's gambit. "In the aftermath of the McCarthy era," Robert G. Lambert has noted, "Arthur Miller felt he could best continue to register his protest by 'adapting' *An Enemy of the People* upon completing *The Crucible*" (628). Conventional wisdom in the literature of secondary education notably links Miller's composition of *The Crucible* with his protest of McCarthy. But Miller's adaptation of the Ibsen play, a play more easy to unyoke from its specific place and culture, has greater universal appeal and force, greater staying power, than Miller's more famous 1953 play *The Crucible,* whose setting in Puritan America blunts its relevance as political commentary. The Preface to Miller's adaptation of Ibsen's play makes only an oblique allusion to the McCarthy era: "Simply, [the play's central theme] is the question of whether the democratic guarantees protecting political minorities ought to be set aside in time of crisis" (8). They ought not to be set aside in crises, Miller believed, and his Preface makes that clear. With the recent passage of H.R. 347, the "Federal Restricted Buildings and Grounds Improvement Act of 2011," the U.S. Government has shown it still likes to try to curtail civil rights (ACLU).

Interpretive cruxes regarding Ibsen's play include the circumstances of its composition, questions about its genre, and misgivings about the character of Dr. Stockmann. Consensus rests on a strongly autobiographical basis for Ibsen's composing *An Enemy of the People.* Emeljanow expressed this consensus with concision: "We know that Ibsen wrote the play as a response to the appalling press he had received as a result of the publication of *Ghosts*" (100–101). As Arthur Miller terms it in the Preface to his own adaptation, "I am told that Ibsen wrote this play as a result of his being practically stoned off the stage for daring to present *Ghosts*" (9). Even if this autobiographical detail is apocryphal, it makes a compelling story nonetheless and could well have appealed to the later playwright's sense of the torment under which apostate artists labor. Generically, too, *Enemy* has caused some disagreement. Neither a comedy nor a tragedy, it has elements of both, qualifying it as a "realistic social problem comedy" (Van Laan 99). For Harold C. Knutson, the play is "a genuine comedy of uncommon merit" (174). For Brian Johnston, the play is a network of classical allusions, as revealed by "Athenian details beneath the imagery of modernism of Ibsen's play" (109). Complication also arises, both aesthetically and generically, from the "Darwinian social determinism" of some of Thomas Stockmann's speeches, which producer Beerbohm Tree found necessary to delete from his working script (Emeljanow 108). Does the attested polemical dimension of the play add or detract? It is vital and enduring, many agree. "I suspect the play will remain as contemporary as tomorrow's headlines, because it exemplifies a commitment to life and to social action that most modern playwrights are unable to make any more" (Lambert 628). On the contrary, others have argued, "*Enemy* is very much in the nature of a dramatized debate" (Adler 87). Perhaps most pertinent in all of these scholarly discriminations, and certainly most pertinent to *Enemy*'s inclusion in the canon of environmental literature, is the public reception of the character of Dr. Stockmann.

Many readers have found an elitist strain in the speeches of this character vexing. The vexation has deepened into grief with the repeated allegation that "'Dr. Stockmann's interpretation of events is Ibsen's'"

(Van Laan 96). The allegation derives from Ibsen's own glib assertion that "'Dr. Stockmann and I get along so splendidly with one another; we are so much in agreement on so many things; but the Doctor is more muddle-headed than I am." (102). Arthur Miller edited out some of the offending elitist passages, realizing that his fellow liberals found them objectionable, even fascist and racist. On those "potentially problematic qualities in the protagonist" (96), though, scholars seem to agree that "Miller has omitted most of them from his adaptation" (96). It can even be said that Thomas's "illiberal" ideas and weaknesses "prevent the drama from degenerating into a theatrical tract" (97). In the wake of Hitler, Miller chose to blunt the impact of those passages, and thereby strengthen Stockmann's character, for he believed Ibsen "would have removed" the extreme speeches of Stockmann in the aftermath of the Nazi Holocaust and rhetoric (100). By subjecting Ibsen to ruthless pruning, most critics contend, Miller did the play and playwright both a service. "Essentially, Miller removes what he saw as potentially a protofascist strain in Stockmann's espousal of an evolving aristocracy of leaders with broad powers to mould community standards" (Adler 87). This pruning, these alterations, prove insignificant in the larger scope of the play: "Apart from the shadings in the protagonist's character and Miller's introduction of more colloquial language, the alterations between original and adaptation might be accounted minimal" (87). Here follow the context for these scholarly disagreements, a summary of the pertinent scene, and a random sampling of Dr. Stockmann's offending lines.

At the height of his passion, reviled by his townspeople, silenced during a town meeting he has called to try to explain the scientific basis for his judgment on the quality of the water, the doctor extemporizes on the nature of knowledge and authority. Insofar as those lines survived the selective editing of the Miller adaptation, they remain alive to denounce Arctic Circle natives in socially stark and Darwinian terms. In the frozen northlands, where Stockmann worked as a physician before getting his job at the health institute, he says,

> Man lives on ice, huddled together in little piles of stones. His whole life consists of grubbing for food. Nothing more. He can barely speak his own language. And it came to me one day that it was romantic and sentimental for a man of my education to be tending these people. They had not yet reached the stage where they needed a doctor. If the truth were to be told, a veterinary would be more in order. (93)

A few lines later, Thomas Stockmann tries to respond to shouted objections and to clarify his view: "Just because there is a mass of organisms with a human shape, they do not automatically become a People. That honor has to be earned!" (93). An exculpatory note for this clearly troubling strain in Stockmann's speech may be sounded by remembering the rudimentary state of anthropology and the social sciences at that time.

The "stage theory" held that "the classical peoples were 'once as much Barbarians as our Indians now are'" (Lemay 204). Stockmann thought he saw, in the ignorant behavior of his townspeople, in their avarice and willingness to be duped by those with vested interests that held the reins of power, a strain of backwardness and illegitimacy. Stockmann is a complex character, complete with flaws. The consensus is right that Miller's adaptation cleanses Stockmann's character and the play of most allegations of elitism or fascism.

More central to ecocriticism, the Arthur Miller of 1950 hinted in his Preface to *An Enemy of the People* at green themes he would not amplify for forty-nine years. "The play is the story," Miller wrote, "of a scientist who discovers an evil and, innocently believing that he has done a service to humanity, expects that he will at least be thanked" (9). "As Stockmann's enemies mount in number," Lambert wrote of the conflict, "Ibsen questions the underpinnings of democracy by asking: what does one do in a democracy when the majority is wrong—totally, incontrovertibly, unquestionably wrong?" (628). Critiques of the Ibsen play have inadequately considered Dr. Stockmann's scientific credentials, although those credentials add to the play's modernity and universality. Stockmann recalls, with exasperated irony, the faulty plan for developing Kirsten Springs: "When they started the damned thing I told them not to build it down there! But who am I, a mere scientist, to tell politicians where to build a health institute!" (36). Stockmann has the training, unlike the rest of the community, to discern ecological truths. He is special. Scientists that work in public-policy circles today, as Todd Wilkinson has shown in the book *Science Under Siege*, share his exasperation. Arthur Miller wrote, concerning Thomas Stockmann and his complicated character, "He was an individualist sometimes to the point of anarchism, and in such a man there is too explosive a need for self-expression to permit him to conform to any rigid ideology" (10–11). That anarchism ties Thomas Stockmann to today's radical environmentalists, those who participate in loose-knit activist outfits and share his antipathy for formal organizations and for laws (Manes; Scarce). But the present discussion need not veer into radical environmentalism to link *Enemy* with contemporary praxis. Arthur Miller acknowledged clearly, in an article written for *Index on Censorship*, that the relevance of the play had changed or grown for him, and that the play might even have outgrown Ibsen himself:

> the story of *Enemy* is far more applicable to our nature-despoiling societies than to even turn-of-the-century capitalism, untrammelled and raw as Ibsen knew it to be. The churning up of pristine forests, valleys and fields for minerals and the rights of way of the expanding rail systems is child's play compared to some of our vast depredations, our atomic contamination and oil spills, to say nothing of the tainting of our food supply by carcinogenic chemicals. (74)

Miller's emphasis shifted, from the "democratic guarantees" and individual freedoms of 1950, to ecological depredations and contamination in his 1989 commentary on the play.

Underscoring this shift in emphasis was a different audience for his commentary on the Ibsen play. *Index on Censorship* is a bimonthly magazine that examines issues of free speech and press censorship. Arthur Miller's commentary, entitled "Ibsen's Warning," expressly highlighted the environmental relevance of the play. Miller qualified his shifting or expanding emphasis on the play (and thereby accommodated his variant readership) as follows:

> It must be remembered, however, that for Ibsen the poisoning of the public water supply by mendacious and greedy interests was only the occasion of *An Enemy of the People* and is not, strictly speaking, its theme. That, of course, concerns the crushing of the dissenting spirit by the majority, and the right and obligation of such a spirit to exist at all. That he thought to link this moral struggle with the preservation of nature is perhaps not accidental. (74)

The poisoned water supply, whether it is an occasion or a theme, becomes a site for contesting factions. How that contest plays out, what sorts of arguments and objections, threats and vilifications constitute that contest, have become part and parcel of our environmental politics today.

The proper study of environment, far more than science alone, must always integrate matters of culture and politics. If, as Adler notes, the play runs perilously close to being little more than "a dramatized debate" (87), then Ibsen and Miller both understood the terms of that debate to be an essence of the play. How did Miller come to conclude, after almost five decades of contemplation, that a central tenet of *Enemy* pivots on nature and nature's economy? How did Miller come to identify, and regard so highly, Ibsen's ability "to link this moral struggle with the preservation of nature"? "Ibsen needed an absolute good for evil to work against," Miller asserted, "an unarguably worthy brightness for dark mendacity to threaten, and perhaps nature alone could offer him that. And, of course, this is even more effective in our time when people have to go to the supermarket to buy clean water" (74). To consider nature "an absolute good" in this context is to shift the locus of the morality radically away from human beings (who are morally flawed for most observers, except for diehard humanists) and to locate such morality instead within ecology, the very processes that give us life and being. The study of the humanities may not be conducted apart from the biological processes that nourish us and allow ours to be Earth's dominant species. Miller ended his 1989 *Index on Censorship* piece with a militant note. In ways that he did not foresee when "The anti-Communist gale was blowing hard" in the 1950s (74), Miller saw in 1989 that Ibsen's drama "literally created a truth as substantial as a sword for later generations to wield against their own oppression" (75). As some observers have noticed since the fall of the Berlin Wall, key members of the radical Left have turned attentions to the environment and to art. Art has the capacity simultaneously to uplift the human spirit and function as a powerful weapon, just as environment often functions as a screen on which we socially project our fondest desires and greatest fears.

To read *An Enemy of the People* as environmental literature, I connect key passages from the play with the environmental struggle today, more than 130 years after its publication. My reading of Ibsen's play, like Arthur Miller's parallel reading of it from 1989, treats it as a central statement of the forces confronting environmental ethics in the twenty-first century. Magistrate Peter Stockmann, a true believer in the correctness of capitalism, reveals no sense of self-irony. He professes "to admire the spirit of tolerance in this town. It's magnificent. Just don't forget that we have it because we all believe in the same thing; it brings us together" (21). Here, ironically, tolerance is characterized by defining its opposite; thereby he foreshadows the later fall from favor of Dr. Thomas Stockmann. Ibsen's depiction of Peter is accurate—and damning. Peter can not understand why all the townspeople are not as enthused as he by the promising local economy. "Everything is shooting ahead—real estate going up, money changing hands every hour, business humming—" (21). One imagines him rubbing his hands in naïve glee. Nor is he aware of the mercenary paradox of his forecast for the health springs: "Give us a really good summer, and the sick people will be coming here in carloads" (21). A "good summer" for his business equates to a bad summer for human health. Illness, that is, equals profits in Peter's paradigm. Corporate spokespeople today, called similarly to account for failing to safeguard human health, decry prohibitive costs and regulatory excesses. And on the rare occasions when questions regarding accountability land in court, businessmen often prevail, for U.S. law accords corporations the same rights as human individuals, all the more today since the 2010 Supreme Court decision, *Citizens United v. Federal Elections Commission*. Early in Act One, Thomas foreshadows the brothers' coming clash when he praises the virtue of "things to work and fight for, Peter! Without that you're dead" (24). Thomas's slogans presage some of the bromides commonly uttered in activist circles today, for example, "If you don't stand for something, you will fall for anything." Reducing complex philosophies to bumper-sticker bites is demeaning for the movement. Neither is Peter Stockmann exempt from documenting his unexamined life. Asked if he does not think his brother should share "a new idea" with the public, he vehemently asserts, "The public doesn't need new ideas—the public is much better off with old ideas" (55). New ideas in ecology today take many forms, many of them progressive and positive—as for example the Environmental Protection Agency's decision in the year 2000 that airborne particulates of 2.5 microns rather than 10 microns are hazardous to human respiratory systems and health.[3] The lowered threshold will save people's lives.

The sloganeering of Dr. Stockmann results in unduly harsh judgments that endear him to few of his townspeople and alienate him from many. Describing to his daughter, to his wife, and to Hovstad the source of the pollution he has discovered—which flows from the tannery of his father-in-law, Morten Kiil—Thomas characterizes it this way: "You know the filth up in Windmill Valley? That stuff that has such a stinking smell? It came down from the tannery up there, and the same damn poisonous mess comes right out into the blessed, miraculous water we're supposed to *cure* people with!" (34–35). Modern audiences will have difficulty appreciating how profane many such utterances sounded to Victorian-era ears. The word *damn* still bore the weight of an actual curse rather than a mere oath, a consigning of the recipient to Hell, and *blessed* was to be reserved for matters of serious devotion. Organizers and activists in the latter twentieth and twenty-first centuries have learned the lesson Thomas Stockmann has yet to heed: irrationality and anger, condemnations and recriminations, are counter-productive for persuading anyone except select audiences in environmental crusades. Such language instead hardens the opposition and disinclines them to compromise, capitulate, or relent. The doctor's language shifts with his shifting moods, and later in the play he lashes his brother with charges meant to wound: "We are getting fat by peddling filth and corruption to innocent people!" (58). This "filth," this "damn poisonous mess," he proves can cause "typhoid and gastric disturbances" (35), which are serious consequences irrespective of the language that frames them. Thomas reaches these conclusions after he "sent samples of the water to the University for an exact chemical analysis." In certain circles today, university scientists are considered suspect. Consequently, the environmental science establishment that tenures them comes under attack (Helvarg 280). Two books from the last decade of the last millennia treat this precise subject, each taking opposite sides in this debate, each demonstrating in its separate way the politicization of science, and each entitled *Science under Siege* (Fumento; Wilkinson). Recurring to more tempered scientific language, Dr. Stockmann reiterates his finding objectively: "It proves the existence of infectious organic matter in the water" (35). His tone changes accordingly, but that change is too late to mitigate the stridency of his earlier statements.

If race, class, and gender are socially constructed today, as culture critics attest, so too is science, both by contemporary environmental politics today and in this play. Disinclined to credit his son-in-law's discoveries, Morten Kiil believes the doctor is exaggerating. So he teases Thomas for trying "to pull your own brother's leg like this" (41). Much like the deniers of global climate change today, Kiil believes the microbes are a practical joke, an invention. "How does it go now, let me get this straight. There's some kind of—like cockroaches in the waterpipes—" The doctor thinks his own leg is being pulled then, and he tries to set the tannery owner straight, but to no avail. "Well, some kind of little animals," Kiil winks. Still today the non-scientist who privileges economic progress over human and ecosystem health speaks skeptically of things that he cannot apprehend empirically, of discoveries that jeopardize the bottom line. "And nobody can see them but you, is that it?" Kiil winks to keep from wincing, because he can foresee what this discovery will do to his business and to his reputation if it proves true. Much later in the play, his back against the wall, he buys up all the nearly worthless stock in the health springs and tries to blackmail the doctor into recanting his dire conclusions. Kiil asks him, "And did you ever think of some disinfectant?" (114). The doctor rightly rejects this futile suggestion as impossible, as a remedy that cannot be accomplished on such a vast scale. "Everything can be killed," the businessman bellows. "That's science!" He fails to distinguish between science (an investigative and descriptive method) and technology (applied science). In the understanding of the greedy and naïve industrialist Kiil, science typically and most usefully deals in death. Typifying the technocratic mind, he yearns to discover or invent a technology that might be able to remedy a problem that technology itself brought on. Morten Kiil hopes to wash clean with a technological disinfectant the damage his tannery has caused. Significantly, that tannery extracts from nature in two ways—both by converting animal skins into human clothing and shoes, then later by flushing the effluent of those conversions into the nearby stream.

The countervailing rhetorics, of revolutionary activism on the one hand and timid moderation on the other, inform the play in interesting ways. This feature of the play also makes it relevant to contemporary environmental conflicts. Newspaper editor Hovstad ultimately turns tail and retreats into his safe economic haven when threatened by Mayor Peter Stockmann. His loss of courage is impressive, because in an earlier scene Hovstad had beaten a revolutionary drum upon hearing the doctor's findings. "Doctor," Hovstad exults, "when I took over the *People's Messenger* I swore I'd blow that smug cabal of old, stubborn, self-satisfied fogies to bits. This is the story that can do it" (45). The story to which he is referring is the story of the tainted water. Miller's stage directions tell us Hovstad is "a graduate of the peasantry struggling with a terrible conflict. For while he hates authority and wealth, he cannot bring himself to cast off a certain desire to partake of them" (20). Miller's directions and notes—one of several ways he "subjected Ibsen's 1882 play to his own interpretation" (Adler 87)—beg for analysis and evaluation. At this juncture in the play they etch in full relief the disparity between the editor and the newspaper publisher, Aslaksen. A flat character, Aslaksen appeals to moderation and temperance so many times that he comically relieves the play from violent threats. He warns Hovstad, "I don't want any wild-eyed radicalism on this thing. I've had enough of that in my time, and no good ever comes of it" (48). Today as well, in environmental debates, the terms *radical* and *moderate* by turns delineate and polarize complex issues. Such terms also foreshadow the ad hominem attacks that so often bedevil efforts toward civil discourse over environmental issues and interpretation of scientific findings.

Enemy is a play, in Van Laan's incisive comment, about "an idealistic outsider at odds with entrenched interests, middle class conservatism, and a self-serving provincial press" (104). For Americans involved in grassroots environmental efforts, being thwarted by the provincial presses is very real today. Hovstad, the outspoken agent of justice and change at the beginning of the drama, renounces his earlier statements and becomes self-serving, "a shill for resource extractionists," in the parlance of Todd Wilkinson, who was describing a similar bit of capitulation outside the stage (83). The "entrenched interests," behind which Hovstad and Aslaksen rally, include those who hold their property most dear. "Kirsten Springs are becoming a gold mine for this town," Aslaksen points out, "especially for the property owners, and that is why, in my capacity as chairman of the Property Owners Association—" (47). Still today, the rhetoric of property-rights interests stands at the center of a growing backlash against environmental sustainability. Indeed, property-rights advocates, basing their arguments on arcane legal statutes like adverse possession, have proven culpable in violence against American greens (Helvarg). The persistence of the property-rights theme in *An Enemy of the People* yields one of the greatest lines, uttered by the character Billing, another newspaperman who proves to be a turncoat later in the play. Early on, though, paying lip service to radical rhetoric, Billing challenges Aslaksen's weary property owners' pitch: "Ha! It's always the same. Give a man a little property and the truth can go to hell" (67). In this speech yet another curse, another vehement speaker, another reductive utterance relate this timeless drama to players in the struggle to conserve and sustain natural resources and human health today.

If whistleblowers the likes of Karen Silkwood have it tough in organizations today (Kohn 13), losing jobs and credibility and sometimes their lives, the parallel in *An Enemy of the People* takes the shape of the backlash against Thomas Stockmann's vow to publicize his discovery. His brother, magistrate and mayor and chairman of the health institute, warns, "You're an official, you keep your convictions to yourself!" Thomas is dumbfounded, and so Peter elaborates: "As an official, I said. God knows, as a private person that's something else, but as a subordinate employee of the Institute, you have no right to express any convictions or personal opinions about anything connected with policy" (56). By these same guidelines, in newspaper after newspaper around the world, reporters who are perceived to be practicing an advocacy journalism while covering environmental issues end up getting sacked. Only tenure saves professors from similar fates. Nor are the pressures only occupational.

Thomas Stockmann resists—selfishly or courageously, depending on one's point of view—his wife's entreaties to cease his advocacy of the water-quality cause. "Tom, Tom, there's so much injustice in the world! You've simply got to live with it" (60). Her appeal shifts when their two young sons enter the room. "What will happen to them? We've got nothing if you're fired!" (60). Such are the sacrifices that the advocate or activist, by any other name, must suffer to advance a social-justice or an environmental cause. It takes a martyr sometimes to see such work to its end. Dr. Stockmann, in fact, flirts with martyrdom. "Was the majority right when they stood by while Jesus was crucified? *Silence.* Was the majority right when they refused to believe that the earth moved around the sun and let Galileo be driven to his knees like a dog?" (94–95). These lines reveal more than a simple "messiah complex" (Adler 89). They acknowledge Stockmann's kinship with visionaries, spiritual and scientific alike, the dynamic and the deprivation at the heart of ecological struggles.

The contest of ideologies, which constitutes the timelessness of this play, reaches a rhetorical fever pitch in the words of Peter Stockmann, whose economic interests are greater than those of his townspeople. Pounding dramatically on his brother's manuscript, which indicts the quality of the town's profitable water, Peter exclaims, "This is the mad dream of a man who is trying to blow up our way of life! It has nothing to do with reform or science or anything else, but pure and simple destruction!" (74–75). More than a century after the play was published, many resource-dependent communities in the rural West are echoing Peter Stockmann's cry, proving themselves to be witless victims of misinformation spread by corporations that stand to gain financially from weakened laws.

When rhetoric fails to impede his righteous brother, when the fury of all the townspeople at the meeting can not slow Dr. Stockmann's pursuit of the truth, Peter Stockmann resorts to legal threats, backed up by laws that franchise corporations equally with individuals. "If you're figuring on opening another attack from out of town," he warns, "keep this in mind: the morning it's published I'll send out a subpoena for you and begin a prosecution for conspiracy" (109). This is no idle threat today. Environmentalists around the United States have become the targets of Strategic Lawsuits against Public Participation (SLAPP). As David Helvarg has noted, "According to Penelope Canan and George Pring, the two University of Denver professors who first coined the term in 1988, SLAPPs are used by corporations and developers to silence public opposition by dragging people into court with spurious suits" (305). And of course the fact that Peter Stockmann is magistrate and mayor and chairman of the health institute means that his reach is almost unlimited—making for yet another tie to the present day, where revolving doors between lobby groups, legislatures, corporations, and the judiciary make progress on environmental matters difficult to achieve. His brother hopes to intimidate Thomas Stockmann into submission by legal means. Hovstad, having swung to the mayor's side by play's end, underscores the serious threat and allows Ibsen an ironic jab at the press: "The Mayor will prosecute you for conspiracy to destroy a corporation, and without a paper behind you, you will end up in prison" (119). These warnings fall on deaf ears. The renegade scientist proposes to unite "street louts, hookey players" (124) to form a sort of grassroots brigade of insurgents to work with him to spread the word about the water.

Embattled men and women of science today, their work thwarted by bureaucrats and agency hacks, the truth they discover squelched in favor of economic growth, have coined a term for themselves that applies well to Dr. Stockmann: "combat scientists" (Wilkinson 2). They cannot perform their jobs without a fight sometimes. Environmental struggles today concern the repression of science as much as its application. If "the structure of scientific revolutions," in Thomas Kuhn's phrase, assures that truth is always provisional, then the interference of politics with science makes the scientific enterprise all the more tentative. In the same way that politicians and judges like Peter Stockmann discredit scientists and scientific findings, public-relations specialists for corporations behave in kind. Ibsen's play evidences both sorts of repression. So does literary history.

> In 1962, with the completion of her classic book, *Silent Spring*, Rachel Carson alerted the world to the insidious effects of DDT and other harmful biocides. But what few young readers probably realize is that her ground-breaking treatise and literary triumph came at a high personal cost. Almost from the moment her research began, the chemical companies that manufactured DDT and other toxic agents mobilized to discredit Carson's work. (Wilkinson 2)

Through the wisdom of the playwright Henrik Ibsen, ecocritics can learn from Thomas Stockmann's stark ordeal. So can theatergoers and attentive readers. All of us indeed can learn that "the tyranny of repression in the workplace is not solely the device of Third World despots or former communist dictators" (Wilkinson 4). The repression of environmental truths continues to thrive in developed nations as well.

NOTES

1. On the movement toward theme courses and interdisciplinary approaches, see *Greening the College Curriculum*, which urges teachers to heed "the urgent call for environmental literacy" (Ehrenfeld x). No portion of that book, including the Grumbling chapter on language arts, mentions *An Enemy of the People* as an environmental text. This omission is conspicuous in the growing body of ecocriticism altogether.

2. "This version of the play," Miller wrote in its Preface, which is "really in the nature of a new translation into spoken English" (12), has kept *Enemy* in the public eye

and appears to have become something of a standard for secondary-education classrooms (Lambert 68). More precise scholars might find it unacceptable for serious study due first to its omission of certain speeches detailed above by Dr. Stockmann and second to its modernization into the dated slang English of Miller's era. As an instance of the corny idiomatic language, newspaper reporter Hovstad, who fancies himself a radical, excitedly urges Dr. Stockmann into action with this sentence: "Well, what do you say to a little hypodermic for these fence-sitting deadheads?" (49). No commentators declare outright that the text was corrupted by the 1950 adaptation, probably because Arthur Miller is so respected as a playwright, but some imply as much (Adler 87; Van Laan 96). The Miller version is available cinematically on American Playhouse home video (*Enemy*), which includes conversations with Arthur Miller and director Jack O'Brien.

3. Environmental Protection Agency. Federal Register Environmental Documents. "National Primary and Secondary Ambient Air Quality Standards for Particulate Matter." 2000. Web. January 25, 2010.

NINE

Gifts and Misgivings in Place

Trying to teach environmental studies to rural Westerners differs from teaching it in cities on the coast. To adopt an easy formulation, made first by the press in the 2004 elections, one might typify the rural inland as red and the coast as blue. Even though mainstream America romanticizes western values, by making the rural West seem simple and bucolic, teaching here is rarely easy. It can require a fundamental shift to reimagine one's audience. To complicate the educator's task even further, publications for tourists and backpackers idealize rural western landscapes, representing them as commodities for sale, but rarely as places to put down roots and learn to get along well with the neighbors.

As someone who teaches classes in literature and environment in rural Eastern Washington, I ask my students to grapple with such issues. I ask how the "tragedy of the commons" might reference more than natural resources in the quality of the lives we lead. In an essay by that title, Garrett Hardin wrote about a hypothetical common ground where ranchers graze cattle and ought to share in a wish to sustain the resource. Instead, every rancher keeps adding more cows or cow-calf pairs to his share till the pasture grows depleted and collapses. That is the tragedy of the commons: that greed and privilege often exhaust the common good. Jared Diamond, in his book *Collapse*, explores such patterns of exploitation throughout a number of eras and civilizations. I aim to widen the semantic range of "tragedy of the commons" to include more basic elements in the hierarchy of human needs. Safety and space are species of resources, I maintain, less tangible perhaps than browse and grass, but substantial all the same. Some Americans, for example, tied to Old West myths of freedom and power, remain bound to firearms and to their rights to use them. And words, like firearms, can be assertions granted constitutional rights.

BESIDE HIGH WATER

One spring the Spokane River was running high, arrowleaf balsamroot yellowing the shoulders of the road, snowmelt booming twenty thousand cubic feet a second toward its confluence with the Columbia River. My wife pointed out the window of the car I was driving and cried. A cat that looked like Weezie, her timid tom gone missing two days before, was mousing on a hill beside the river. I coasted the decrepit Honda wagon to the gravel shoulder and set the brake. Quite convinced that she had the wrong cat, I never got a chance to learn for sure. Out of the car she leapt and began to call and coax. We made a family tableau—her crouching, me akimbo between the car and open door. A magpie lit on a telephone line, and I turned my attention away from the stooping woman and the cat.

Conditioned by my past life as a hunter, I thought about dropping that magpie from the wire. Magpies, the most voracious avian predators in the West, rob nests and nestlings. They eat crops and ravage gardens. One hunter claimed they like to enlarge wounds in cattle and hop inside to devour them from the inside out. Inside the car lay a powerful pellet rifle. The day before, my son and I had been setting the sights, and I had forgotten to put it away. No one but Karen would witness my little lapse from the law for a good cause. National law now protects magpies, but when I was a kid a bounty rested on their heads. My musing on the bird came to a quick end, though, when a rock slapped into roadside weeds and the magpie flew. Another rock whizzed overhead, this one colliding with a Yield sign some thirty yards away. The clang echoed off the road.

On a stump behind us stood a young David. From one hand dangled a sling, two laces linked by a leather pocket they whirled. What a crude weapon to master and grow so precise. I shouted at him. I swore. He gazed our way, abstracted, as if regarding us for the first time. Then a cloud crossed his brow and he slid the sling into the bag at his hip. I had made him sore, I saw, maybe made a foe of him. Narrowly he eyed our car as if to memorize it. And then he turned and angled along an uphill path away. Karen crouched still behind the open door. The Yield sign was pocked, I saw, where rocks had struck it.

Less angry than anxious, I was trembling and tasting copper. What if he had set out to pelt us? What if my hurled curse or cry had worked him up enough to draw from his fanny pack a pistol he'd held back? Karen, a criminal defense attorney, knows how dangerous some citizens can turn when they get riled up. She often warns me not to anger fellow residents, or to yield to angry drivers in cars. For myself, I have lived long enough in the rural West to have witnessed or heard tell of lapses of judgment with guns. Greedy and illegal hunting out of season, target practice endangering anyone in range, killing for sheer sport. Two off-duty cops in nearby Suncrest, one an officer, the other a new recruit, had gotten drunk and let fly one night with their firearms out the back of the mobile home where one of them lived. A stray slug had entered a house and lodged in a wall near the head of a woman in bed. Their sporting weapons included a strictly illicit automatic.

It's not the firearms alone that disturb me, although plenty of bad gunplay takes place around the West. It is the disregard of some of my fellow Westerners for people, animals, property, and common sense that astounds. Nor could I claim to be free of its symptoms—I who had just been tempted to drop a magpie from the wire. How can so many of us ignore the freedoms and needs of the other beings that surround us?

Following our encounter with the lace-whirler, the reckless thrower of stones, I brooded for some time on what I had seen. Late that night I lay, mired in an old quandary involving rights and liabilities, privileges and guilt. "In dreams begins responsibility," William Butler Yeats wrote long ago, and I knew what he meant. Freedom and peace necessitate a weighty set of obligations. George Lakoff characterizes such obligations as "responsibility, which is at the heart of liberal/progressive morality." Such morality "begins with empathy, the ability to understand others and feel what they feel" (62). Still today I am learning what motivates my red-state neighbors within blue-state Washington.

THIS GEOGRAPHY

My particular bioregion, the location of the main campus of Eastern Washington University, lies at the edge of three ecotones, three transitional areas between adjacent ecological communities. To the north and the east, the topography rises and gives way to stands of coniferous forests. Mountains loom there, staggering canyons, swift rivers, and enormous lakes named Priest, Pend Oreille, and Coeur d'Alene. To the south ranges the Palouse, one of the planet's most productive wheat- and lentil-growing regions, with its steep rolling hills and deep richness of volcanic loess soil. To the sere West lie channeled scablands and shrub-steppe, essentially desert, relieved in spots by potholes and seeps. The word *channeled* denotes the remnant paths of prehistoric floods that coursed across volcanic basalt soil and rock, while *shrub-steppe* names the arid scab-rock lowlands and treeless scrub brush threaded by badger burrows, shallow soil, and rocks. This region, encompassing Northern Idaho and Eastern Washington, has a political identity commonly called the Inland Empire, a place dependent for its economic stability on immemorial natural resources: timber, mining, ranching, agriculture, and fishing by Indian tribes.

Extracting those resources into perpetuity is not desirable or possible, for they are finite. The latest culture war, since the spotted owl clash of the 1990s, concerns flagging salmon populations in the Snake and Columbia River system. Dams erected a half-century ago have come to function less as flood control and irrigation than as ports for barging crops and paper pulp the hundreds of river miles to the sea. Salmon restoration, as mandated by the Endangered Species Act, costs a billion dollars a year amid business as usual. The Indian tribes—Colville, Kalispell, Kootenai, Nez Perce, Shoshone, and Spokane, to name a few—enjoy adequate legal standing to press for restoration of the fish, but they continue to hold back. That battle is only one tense aspect of my place.

Just as important for my students are the eroding farm fields, the soil banks failing due in part to chemical inputs and unsustainable patterns of use that trigger tons of topsoil to be spirited away. Such cultural transformations can teach college students something. This region is relying more on technology, recreation, manufacturing and tourism than on natural resources. As those same natural resources dwindle, the qualities of life erode, and our dwindling natural resources come to be more precious than jobs or gold.

LIGHTING OUT FOR THE TERRITORIES

One traditional trait of the western provinces has been free reprieve from social obligation, a respite from moral codes. One may think of Edward Abbey, who moved from Pennsylvania to New Mexico, studied European anarchy in graduate school, then settled in southern Arizona where he began to commend radical activism, including "monkey-wrenching" or sabotage, and helped to found Earth First! Abbey's strain of activist environmentalism was essentially anarchical. The literary legend Huckleberry Finn is aiming likewise to "light out for the territories" at the end of this book, to head west and flee the domineering authority of the prim Miss Watson.

Many of my neighbors would appreciate Huck's longing. They are just as prone as Huck to mobility and itinerancy, as skittish of obligation and blame. Accountability to others and to place—an aspect of the bioregionalism I advocate in class and practice in person—is hard to champion when so many forces militate against it. Inland Northwest residents do not like others telling them how to live, especially if those others represent the government, whose proxy as a state employee I am. They will not be told what to do.

Their irascibility is an aspect of place. *Winter Range*, a novel by Claire Davis who lives in the nearby Palouse bioregion, centers on such tensions of mutual accountability. When a renegade rancher goes bankrupt one winter and cannot afford to feed his stock, he spurns those who would dispose of his animals humanely, allowing no one to feed them, buy them or give them care. In a death spiral of a plot, the rancher starves his animals, starves himself, and sets fire to his banker's home, effectively punishing his community for an economy in transition and for his shortsighted failure to manage his own ranch. In *Close Range: Wyoming Stories*, the collection containing "Brokeback Mountain," Annie Proulx explores the lives of cockeyed sodbusters and limp-alongs who are the living equivalents of a middle finger lofted high. "Indigenous" in these parts is as apt to mean "lawless" as it is to mean "native-born." Vigilante justice holds sway in remoter regions of the rural West, as the Randy Weaver story shows. After selling a sawed-off shotgun to a federal agent, Weaver refused to appear in court. He held out, holed up, found his home under siege, his son wounded, his wife dead on his cabin floor (Walter). He and his life story are aspects of my place.

If renegade survivalists stockpile food against Armageddon, instructors at nearby Fairchild Air Force Base teach survival methods to prepare pilots for the downing of their planes. Some college lecturers feel like crashed travelers, uprooted from family and place by circumstances, plunked down in strange lands, chasing short-term jobs. Sheer survival as priority, the mere endurance of circumstances, can sometimes make the higher aims of self-actualization go dormant. People cannot fulfill their hierarchies of higher needs when they feel unsafe in their environs. How can anyone care enough about a particular place to gain its intimate acquaintance when the personal needs have long gone begging?

Even within its urban concentrations, though, the American West can stifle. How well I recall a pair of African Americans who located to the University of Idaho to play football and enrolled in a writing class I taught. When I tried to do some environmental education, they were skittish about being outdoors. They high-stepped through the tall grass of the placid university arboretum. I could guess why. They had heard about the Aryan Nations headquarters, the compound less than an hour away, owned and run by the Reverend Richard Butler, he whose white supremacist ideology had dubbed black folks "mud people." Once the Christian Identity movement came to impinge upon my consciousness of place, mentions of "identity politics" took on a new edge. To teach about my particular place with honesty today means to factor in that Christian Identity.

Once I moved to the inland West to stay, I had to answer to friends in Bellingham and Seattle who wondered why I had chosen to light out for these territories. I said, and I maintain it still today, that the forces of poverty, ignorance, and fundamentalism can best be countered here. If one aims to effect social change, the territories are the place to be.

During my four years as a visiting professor at the University of Idaho, I designed a course entitled "Issues in White Supremacy" for the Martin Institute for Peace Studies and Conflict Resolution. Attorney Salmon O. Levinson established the William Edgar Borah Outlawry of War Foundation at the university in 1929. Since 1948, the foundation has sponsored a yearly program devoted to understanding war's causes and the conditions needed to establish lasting peace. How ironic it proved when the Aryan Nations set up shop just north of there in the 1980s. Teaching my course in white supremacy helped me to learn as much about my adoptive homeland, certainly, as the students learned from me.

IN THE CLASSROOM

Environmental educators agree on the need to foster sensitivity to place inside and outside the classroom. Students may gain empirical grasp of the fact that we are animals first—mammals who share 98 percent of our genetic material with chimpanzees—and humans only second. In the primary grades particularly, but also in entry-level college classes, students who are allowed to explore and develop connections to nature in their personal lives appear from research to be more apt to thrive as citizens and professionals. Wisdom deriving from experiences, emotions, memories, and personal histories can encourage a virtual reinterpretation of the world. It can cultivate interdisciplinary sophistication and counterweight the faulty warnings toward objectivity that unbalance the scales inside so many disciplinary forms of knowledge. On the most intimate levels, our place-based attachments to the past can enlighten and empower our students to shape the future.

My students join me on field excursions, but I also ask them to excavate their personal pasts. The excursions take place typically in the Turnbull National Wildlife Refuge where my university maintains a field station. We hike and listen, learn some flowers and birds, all of this a preparation for a classroom component on place-based attachments. There is no way I could ever presume to choose a place for my students. I can only show them the ropes and rewards of visiting the refuge wetlands. To help them own the experience, we spend some time identifying favorite places from their past and present, and then I ask them to compose three separate essays based upon three exercises.

First I ask them to imagine their special place from the perspective of someone other than themselves. Project themselves, for instance, into the mind of an animal, bird, or tree they have seen; a character, narrator, or author they have read; or a farmer, rancher, or logger they have met. They write freely about their places using the voice of this person, entity, or being as they imaginatively conceive that voice to be.

Second, I ask them to analyze their geographical place by separating it into some of its component parts—the air, water, soil, and the variety of lives they maintain. Alternatively, they might study how the parts of the place fit together. How do they rely upon each other? How do they integrate to make a whole? They narrow the analysis then to some human impact. For instance, how does internal combustion affect their place?

Third, they judge the value of their place according to a single principle. What value, for instance, does the place hold as bird or animal habitat? As a source of beauty? For human recreation? How about ecosystem services, like water filtration? These exercises get them thinking about the value to their lives of special places.

A decade ago I was teaching English 308: Advanced Exposition, a themed class cross-listed with Honors. That class entailed field excursions, environmental and social-justice issues, and techniques of civic writing that can effect social change. Two older students arrived with chips on shoulders and pens in hand to record transgressions they were sure they would find within the liberal professoriate. I thought to tell their story when I heard some news about an alumni organization that was paying UCLA students to spy on professors. Students were lured, in exchange for recordings of allegedly partisan lectures, by the opportunity to earn $100 a pop for their reports (U.S. University).

Mine was a small class, fewer than fifteen students, most of them women. Two days a week it met, first in a windowless room with clusters of tables. Later we switched to a fishbowl lounge, with windows and doors on three sides. The peer-group work for that class necessitated a level of trust and vulnerability, a relinquishing of anxieties about one's writing and about the personal disclosures the assignments often prompt. In ways we do not fully understand, environment shapes identities, or so research by Thomashow has shown. Memories of beloved places, degraded places, and lost causes are common when I set out to assist students to recover memories of beloved places from childhood. Poet William Wordsworth's legendary "emotion recollected in tranquility" was as much a component of his poetic project as it is for my students excavating their own pasts.

As with many classes at my regional university, this one had a number of non-traditional students. There was Martha, the blue-haired grandmother, active in family and church, aiming to train as a travel writer. There was Fred, newly divorced and crew-cut, retired from the military, who cradled a motorcycle helmet, a state-required accouterment for his "freedom machine," as he named it. There was Nicky, married mother of two, in her late twenties, open-minded but quiet and shy. And there was wanda (she did not capitalize her first name or her surname), an African American single mother who was recently discharged from the Navy and sported an eyebrow ring. She was outspoken, savvy, and cocksure. The other members of the class were more traditional-aged students from a variety of majors, most of them nearing the ends of their college years, free thinkers who found the class praxis attractive—the field excursions, the environmental and social-justice studies, and the techniques of civic writing I developed as the core of the course.

Larry was a retired "officer and gentleman," as he dubbed himself in a series of emails, a military man who had spent decades in the service. He had come to this region first when he got stationed at Fairchild Air Force Base in nearby Airway Heights. He disclosed also that he had lost a job as an accountant, which occasioned his attending college in his fifties. Much later I learned he was a student of Ayn Rand and a tax resister. He was probably a libertarian. He had addressed the John Birch Society, whose local branch rents billboards to seek converts for its opposition to U.S. membership in the United Nations. President George W. Bush, we all might recall, had refused to cough up U.N. dues. He would later name a U.N. Ambassador who had publicly opposed the United Nations.

So I found myself heading a small class, grounded in study of activism and place, a class that contained three retired military people, two of them white males in their fifties, one of them a black female in her twenties. Larry objected vociferously to the assignment for service learning, which I defined as learning that is altruistic. He protested that assigning altruism isn't right, it isn't fair, and it does not constitute sophisticated learning. I tried an example, a tentative assertion that some people turn to teaching as an unselfish endeavor. He scoffed and said teachers get paid far too much for teaching to be named a species of altruism. That was the first day of class, and I knew we were in for a wild ride.

On the second day of class Larry decried examples of environmental restoration. He ridiculed our coming field excursions as "bunny worship" and "tree-hugging." When I handed out poems by Wendell Berry and Mary Oliver, and asked for volunteers to read them aloud, a pall hung over the classroom. A large man, well over two hundred pounds and six feet, Larry was intimidating them. In an email he sent me the following week, after I had laid out some ground rules regarding civility, he chided, "This is a junior-level college course, not a tea party with polite little old ladies attending to sip some tea, have cookies, and then perhaps play some bridge. Civility is a wonderful thing, especially at tea parties—and elsewhere, if and when possible." He was merely exercising his right to free speech, he reminded me with a virtual sneer, appealing to the civil libertarian in me.

About midterm, the situation grew dire during a classroom discussion. We were speaking of Irish slurs as used by Frank McCourt in *Angela's Ashes*. The talk turned to other slurs. wanda said that she considered the preponderant "n" word offensive in the writings of Mark Twain. Eloquently she explained that African Americans read history and literature differently from white folks, due to the centuries of oppression and slavery they bring to it. Reader-response theories of literature support her, I said. Larry rose up in objection and accused her of advocating censorship. In a heated exchange with wanda, Larry used the offending word repeatedly, claiming that his honest aim to contextualize the word's use in American history cleansed it for classroom use. His righteous insistence on free speech was deliberately pushing the boundaries of civility I had established in my ground rules for the class. The entire educational atmosphere had taken a libertarian tilt. The world was looking cockeyed, and objects were beginning to slide off tables.

During the following week of class, wanda and Nicky stayed away. They wanted to learn to write, not debate civil liberties, they told the dean of students. I consulted with the dean and initiated action that banned Larry from the class. His emails to my chairman and me escalated, ever more strident and menacing. In an odd twist, he tried to claim the peculiar decorum of Congress and Parliament for our class. There, "during floor debate you see people at times uncomfortable, intense and even adversarial." Larry was wrong, of course, his reasoning fallacious, to draw an analogy between a public institution of higher learning and the behaviors demonstrated by elected members of legislative bodies. He had hijacked our class and made it his private laboratory for the First Amendment.

Larry thought he was working from within our system to reform it. "You should get your ass fired," he wrote, "and I sincerely hope you do because of your incredible dishonesty. If someone has not 'felt safe' in this class, then they might have some personal problems to work out. If someone has felt 'uncomfortable' or 'hurt,' then perhaps these individuals should stick with the tea parties and women's socials." In later correspondence he addressed the lamentable state of "post-Christian America." I allowed him to complete the class, and for a paper he was drafting I invited him to read and reference Pat Robertson's *New World Order*.

Larry took to lurking and asking his erstwhile classmates to accept unsolicited information on free speech. Nicky and wanda reported to the dean they were feeling trapped. Following their report to him, the dean put teeth into the suspension and ordered Larry to stay out of the building and have no contact with his classmates. Strangely, I felt guilty about this last-resort maneuver of the dean's; until I met Larry, I had been an unrepentant advocate of free speech. He complied with the order but began to threaten to sue me for slander and to challenge whatever grade I chose to give him in the class. Happily for the students, the cloud of threats evaporated when he finally went away.

Larry's behavior began to flash me back to the decade before, when members of the Aryan Nations had leafleted parked cars at the University of Idaho campus. Racists from outside the campus were aiming for recruits from the student body. At the 1993 meeting of the Rocky Mountain Modern Language Association in Coeur d'Alene, Idaho, just down the highway from Hayden Lake, where Richard Butler hung his swastikas and lit ceremonial crosses, I had presented a conference paper. The RMMLA, adopting a theme that year of ethnicity and race, set out to thumb a nose at the racists nearby.

During the keynote address at the conference, four members of the Aryan Nations crashed our meeting. They made their way into the ballroom—dressed like street punks in leather jackets, high boots, patches, insignias, and berets—and stood directly behind my folding chair. It was a tense stretch. They did their best to commandeer the discussion and redirect it toward home schooling. One of the four, Floyd Cochran, defected that next year when his comrades would not support his right to keep alive his "defective" cleft-palate child (Hochschild). They believed the child should be trashed, like a defective part. Something about the purity of the gene pool seemed to be at stake for the Aryans there.

FOR THE TIME BEING

At the same time that I was losing control of my environmental writing class, the Aryan Nations was losing its grasp on the region. The headquarters at Hayden Lake lay northeast from my office at Eastern Washington University some forty miles, while the University of Idaho lay southeast some ninety miles. A car carrying a mother and son, Victoria and Jason Keenan, while passing the ramshackle compound backfired. Mistaking the exhaust report for a shot, the guards chased, assaulted, and fired on the family. In the wake of that event, Morris Dees of the Southern Poverty Law Center saw his chance and sued on behalf of the Keenans. He bankrupted the Aryan Nations. The Reverend Richard Butler had to sell his house and scattered outbuildings to pay damages. He died in 2005. The "world headquarters" of his organization moved to the Midwest.

The territorial imperative the Aryans hoped to foment in lily-white North Idaho proved too slippery to grip, much less to expand to include the western states north of California, including the western Canadian provinces and Alaska. This expansive area was to have been named the Aryan National State. Incidentally it lay along the same lines as Ecotopia, the fictional province based on ecological sustainability and envisioned by Ernest Callenbach in his 1975 novel of that title. Today the West Coast is known as the Left Coast, among some conservatives at least, for its concentration of progressive types.

Something about the far Northwest has attracted radicals for decades. Wobblies, the International Workers of the World, built a short-lived stronghold in Portland, Seattle, Spokane, and Everett in the early decades of the twentieth century. Their heirs today have established some tentative alliances with environmental groups. Robert Mathews in the 1980s split off from the Aryan Nations, founded a group called The Order, and "recruited followers to commit murder and robberies here because Butler wouldn't back his call for violence. The cold-blooded killers of Denver radio talk show host Alan Berg were spurred to action here by incendiary speeches delivered by Butler and his comrades in hate" (Oliveria B8). More recently in the Northwest, the FBI has rounded up one cell in the shadowy group named the Earth Liberation Front, whose participants have inflicted some fifty million dollars in economic sabotage to labs and housing developments in the American West, a story told well in the recent documentary film *If a Tree Falls*.

BURROWING IN THE BACKYARD

Despite an occasional sour experience with a renegade student like Larry, teaching has its immaterial rewards. In a different ecocomposition class I taught several years ago, here is the process one of my students followed to explore and establish her sense of place and ecological identity. She did not know it at the time, but she learned in the process to practice a bioregional ethic. Peter Berg, who originated the concept of bioregionalism, defines it this way: "The concept of a bioregion as the basic location where people live, and the practice of reinhabitation of that life-place by its residents, are necessary to rejoin human beings into the overall web of life. Harmonizing with the natural systems of each bioregion is a necessary step toward preserving the whole biosphere." My devotion to my Pacific Northwest homeland affected Hailey and helped make her work more meaningful.

Just south of my university is a bioregion, the Palouse, where sophomore Hailey Booth had lived her entire life. This area of fertile soil bridges two watersheds, one west to the Columbia Plateau, the other dropping south to the confluence of the Snake and Clearwater Rivers, both draining ultimately to the Columbia River and west to the Pacific Ocean. A breed of horse, the Appaloosa, was named for its genesis on the Palouse, a fact Hailey learned only after starting to study this place she had always considered her own.

One autumn afternoon, Hailey saw the autumn sun grow dark by 2:00 pm. Before her saintly neighbor could consult the Book of Revelation, drops of mud began to rain upon his Chevrolet. How in the world did mud come to fall from the sky? Only much later did Hailey learn that the mud was made when rain passed through clouds of airborne dirt and bore it earthward. She witnessed that marvel firsthand and recalled it in a journal entry written for the class: "The sky was purple. Mr. Smith stood in his yard and gazed at his new car. By the time he turned around, his white dress shirt was splotched raw brown."

That is how her research began in a place she thought she knew. Some 20,000 years ago when the glacial ice dam forming Lake Missoula broke and broke again, lake water rushed through to the north and to the west of the Palouse, scouring the topsoil and weaving a region of coulees and channeled scablands—a land rich in rivers for spawning salmon, until hydropower projects like the Ice Harbor Dam ground the fish runs to a halt. The Palouse soil is mostly loess, she discovered. In geographers' jargon loess is "a buff to yellowish brown loamy deposition of North

America and Asia," easily airborne, which appears from the records to have been blown in after settling from the rushing waters in adjacent bioregions and collecting in a second prehistoric lake that eventually went dry.

Hailey also discovered the names for vegetation indigenous to the Palouse: camas and bunchgrass, lupine and paintbrush, Rocky Mountain penstemon and ponderosa pine. Such species created biological diversity, which she defined as "variety in habitat and life forms." This biodiversity encouraged animal species now generally shoved out: grizzly bear and wolverines, bull trout and chinook salmon, sharp-tailed and sage grouse. Native gamagrasses also often have been supplanted by exotics like cheatgrass, starthistle, and knapweed. Commercial crops, wheat and peas and lentils, now blanket the steep hills.

To learn what forces impacted the indigenous species, Hailey chose to study the short history of farming. The Palouse hills possessed an astounding 100 to 200 feet of loess topsoil when her ancestors, hungry for land to plow and plant, migrated in the early nineteenth century with the likes of the Reverend Marcus Whitman. Sodbusters, spurred by Manifest Destiny, came to be despised by mountain men. The Indians despised them, too, and finally Indians murdered Whitman and his wife in 1847. Hailey learned how many acres the Palouse farmers have cleared and how many lie fallow today, subsidized by funds from the Conservation Reserve Program, "a U.S. government contract offered to farmers who own an abundance of farm acreage." Hailey came from a farming lineage, and her own ancestors would have ranked among those the mountain men despised.

I had shared some of my own writing with the class. In an entry for her journal, she helped herself to one of my own poems. "The head-high grasslands must have been more lively than these fields fuzzed green with wheat." Tender-eyed pioneers "compared the sight of grasslands to the sea, to the sky," and to "blowing animal fur" in the beginning. Prairie chickens, grouse, fowls of all species bred and nested on the sod, and "plow" was a sound that seed heads made when air went gushing past. Imagination, a wobbly first step toward the development of critical thinking, did not threaten to overtake Hailey's scholarship in her journal entries. It promised more appropriately to modulate the scholarship. If all work and no play make Jack a dull boy, students of every age benefit from language frolic and imaginative exercises that may trigger critical thinking.

At that stage in our online exchanges, I introduced Hailey to a term I learned in my research: *monoculture*. This term characterizes the way Euroamericans raise grain crops, graze dairy and beef cows, and plant nonnative trees in place of the native trees they cut. Farming systems often work by eliminating "weedy" or "predatory" species and reducing native diversity to single crops. This is a sound system economically, but it can produce biological ruin. What is forfeited with the loss of biodiversity, Hailey learned, is some tangible integrity of place. "To know the spirit of a place is to realize that you are a part of a part and that the whole is made of parts, each of which is whole," the poet Gary Snyder had written (38), and the wholeness of her home place began to seem fragmented. "But what has really been lost?" she asked in her journal as the exchange grew more intense, as the voices of historians and Christian pioneers, combat scientists and farmers and poets came together to weave an interdisciplinary web.

What really has been lost? That question leads to ethics, which is probably the trickiest intellectual region of all. If the loss is not a material loss, is it really a loss at all? Just like that old conundrum about a tree in a forest, if no one misses the things that are gone, does their absence matter? More to her point as an aspiring literary activist, if no one expresses the loss of things peculiar to a certain bioregion, is that loss really a loss at all? And here she began the search to find her place, not a search from room to room but the search within herself to find out how she felt. Is she any the poorer for never having gotten to live with extirpated species, sage grouse, sharp-tailed grouse, salmon and trout?

Further reading about farming revealed that only within the lifetime of her father did the use of chemical "inputs" begin—fertilizers, herbicides, insecticides, fungicides, and fumigants. How many sprayings per year? How many tons of toxins per spraying? What "target species" of weeds and bugs need to be controlled? What "non-target" species get a dose as well? Hailey had read excerpts of *Silent Spring*, Rachel Carson's study of the impacts of DDT in the 1950s and 1960s, as well as Edward Abbey's lament for bats in Carlsbad Caverns diminished by toxic sprays used to regulate crop pests in West Texas.

Back to that autumn day when winds blew hard and the sky grew dark, what caused that all to happen? In 1980, when Hailey was ten, Mount St. Helens had blown its dome and hurled tons of ash across her bioregion. But no volcanoes were erupting when her sky grew dark and raindrops turned to mud. *The Grapes of Wrath* furnished a context. "Tractored out" of Oklahoma, exiled to California, the Joad family of migrant laborers in the novel by John Steinbeck became victims of ruthless banking, careless agriculture, and drought—a perfect storm of conditions, both natural and human-made. The Oklahoma Dust Bowl started in part from ignorant farming practices. Another term, "sustainability," entered the range of debates that would later comprise her research paper sequence for the course. Depression-era farmers did not farm their land sustainably. Without knowing about contour plowing or rest-and-rotation methods, they tilled and planted the Midwest plains in ways that could not be sustained. They exhausted the land. A place they had started farming, a place that was whole, was losing a part of its wholeness every day the wind blew. And as a result the human culture, too, could no longer be whole in that place.

Could it have been topsoil that the Palouse wind lifted above the region? Hailey interviewed agronomy professor John Reaganold of Washington State University and learned that an acre of Palouse cropland loses five to twenty tons of topsoil every year. How does a ton of dirt equate to fractions of an acre-inch? How far can topsoil sail on a good stiff wind? Which streams does it flow into, and how far away? How much topsoil is left today? Questions generated questions. "Do you see those fawn-colored spots on the hilltops?" Reaganold had asked. Those are known as "clay knobs," places where the topsoil has been washed away by rainfall or snow melt, blown away by winds, all one hundred to two hundred feet of topsoil the pioneers discovered and subdued, as the Book of Genesis enjoined them to do. Who says you can't see wind? Hailey had seen the wind and it was brown.

For her second required interview in the research project sequence, Hailey met with Nancy Taylor of the Palouse-Clearwater Environmental Institute. For part of every busy year, Nancy helped to farm her family land in northern Utah. The rest of the time she wrote grants, performed community outreach, and attended meetings of the National Organic Standards Board. Nancy argued—and to Hailey her passion was plain—that our farming practices cause the soil to blow and flow to parts unknown. A mere sixty years ago, no one spread chemicals; farmers rotated crops, fertilized by animal or green manures, let the soil rest or lie fallow, and erosion was never a problem. In the decades since World War II, though, most farmers have become "chemically dependent," a clever original phrase of Hailey's that resonates. Since the war, a powerful techno-industrial economic system has emerged, one with vast implications both for the food we eat and the air we breathe.

While organic farming may not produce such high yields as chemically dependent farming, Nancy demonstrated to Hailey that it promotes biodiversity, exempts the land from sprays, and keeps chemicals from our foods and water tables. Moreover, it slows erosion. Roots and leaves and decaying organic materials, brisk with microbes building compost, enhance soil stability. Soil heavy with organic nutrients would not have been lost to a whim of the wind like the one that Hailey witnessed on that stormy autumn afternoon. "You don't have to be a 'granola' to agree that organic agriculture offers ways to control the problem of soil erosion on the Palouse," Hailey concluded her major researched argument for my writing class. In doing so, she appealed skillfully to a set of assumptions that are apt to be shared by her audience. At the same time, she riffed on "granola," a pejorative moniker still not admitted to many standard dictionaries.

The cloud that she had seen that day, the wind that had obscured the sun, was soil erosion. It began with the tilling of too-high hills, it grew worse with the loss of health the soil suffered when chemicals were sprayed for decades, and it got caught up in the high brisk winds for which the region is known especially in the spring and fall. The research papers Hailey composed that quarter were compelling successes—drawing as they did from interdisciplinary sources, driven as they were by curiosity and a native passion that began in her attachment to her place.

HERE AND NOW

Many of my colleagues adopt an objectivity regarding matters of politics, race, class, gender, economics, and environment. They are aiming for the treasured scholarly disinterest that is believed to characterize good classroom demeanor. Objectivity in such issues is a myth, I believe. No crime or liability ought to attach to outspokenness on issues, and I worry sometimes that my colleagues have lost touch with what it means to profess. At the same time, my students may write whatever they like, no matter how it runs counter to my own philosophies, and I will grade them fairly. Indeed, I will respect them for taking risks, for refuting conventional wisdom and questioning my authority.

Students everywhere should enjoy due process when they think they have been graded or treated unjustly. Students and parents are paying tuition—paying more and more dearly every year—tuition that helps fund the salaries of professors. We ought to be held accountable within education, just as citizens and neighbors ought to be accountable to one another, to other species, and to the sustainability of the water, air, and land.

Environmental studies, the study of ecology and place, ought also to examine our bioregions as much for their social and political inner workings as for their topographical and biological parts. If we do not possess a basic understanding of market pressures and population dynamics, of growth-management mandates and constitutional arguments about eminent domain (Ray), then those of us who teach about the dynamics of place-based attachments will be giving incomplete views of the complexities involved.

Larry eventually followed through with the formal protest of his grade, and an appeals board upheld the low grade I gave him. He never made good his threat to sue me for slander. My attorney-wife tells me the charge carries a very high burden of proof. Still today I sometimes see him riding the city bus between Spokane and Cheney where I work. I look through him and regard the landscape out the window as he scowls my way.

Long ago I lit out for the territories, leaving behind the city of Seattle where I was born. I chose to work in a conservative region, and that has made all the difference. It has its challenges, this teaching about environment and place to occasionally adversarial audiences, and I confess I sometimes wish that I were once more preaching to the choir in the blue-state half of Washington. But I continue to fight the good fight, partly by asking my students to look their places and circumstances in the face. With caution, skepticism, and a dose of hesitancy, I situate difficult people, ideas, and social movements within the landscape. Custom and culture make them parts of the social ecology of the rural West.

TEN

Restoring Bioregions Through Applied Composition

"The language he speaks and writes is spoken
and written by no one but him, which solves,
for him at least, the problem of audience."
—Robert Wrigley, "Babel"

From twin perspectives—as an environmental organizer and a professor—I write this. As an organizer, I lead the 1,200-member Upper Columbia River Group of the Sierra Club, a regional affiliate of the world's largest environmental organization, a group stigmatized in some circles for being bureaucratic and mainstream. The Club plays upon the place-based attachments of its membership, enhancing those attachments through educational outdoor excursions. Our members are conservative in the foundational meaning of the word *conservation*. They are resistant to the changes that cause the destruction of beloved wild places. As a professor I practice service learning and some environmental education, the latter under increasing scrutiny by right-wing think tanks, foundations, and Congress. My research makes me ever mindful of the politicization of the humanities and sciences alike. Indeed, recriminations of higher education (e.g., that it is a last bastion of Marxism and a haven for tenured radicals) often focus on American studies and literary studies. The need I see for an applied composition emerges from a different critique of the profession, though. The distinction between exoteric and esoteric culture helps set up the present set of boundaries for my argument. If exoteric culture encompasses arts and entertainments suitable for communication to the general public, then the esoteric arts are recondite, obscure, and capable of being comprehended only by the select few, the inner circle, those among the chosen who have mastered specialized forms of knowledge.

Much of what you will find written about the language arts demeans good writing. It accords primacy to theory over praxis, and it carries those theoretical dimensions far beyond a healthy balance, diminishing the honest promises of succinct communication. The subject matter of such theory is, as Hamlet objected, "words, words, words" and little more. Words that belie the scrupulous claims of those who compose them; that complicate the problems of race, class, and gender they seek to alleviate; that do nothing to offset the imbalances of power that typify techno-industrial society. It is elitism masquerading as egalitarianism. The problem lies in misperceptions about audience and purpose, the cornerstones of rhetoric. If scholarly fascination with theory and abstraction contradicts the needs of those less fortunate species and people we champion, then cruel ironies will ensue. What is said in jest is largely true—while the deconstructionists were seizing the English departments, Ronald Reagan was seizing the world. Who fell asleep at the watch? Carl Safina noted of graduate study in the sciences, "Oddly, much of my academic training discouraged meaningful involvement in the public arena. Professors taught graduate students to purge from research proposals any taint of the 'applied'" (A80). Safina's observation rings true for the pursuit of advanced degrees in language and literature as well. Various so-called discourse communities maintain their own lexicon. And yet the syntax and diction, the styles by which certain professors of the language arts endeavor to map routes toward an improved society, all too often constitute an "official style," in Richard Lanham's excellent phrase (56–80)—arrogant, specialized, void of particularities, estranged and estranging, even to a majority of colleagues in languages, communications, and literatures. I propose we loosen the grasp of official styles, shrug off careerist pretexts "to sound learned and scientific—disinterested, impersonal, factual" (60). Let us begin instead to legitimize the profession through innovative research and validate personal experience in our work and classrooms.

Language study as skills only, for instructors and students alike, can stultify. It is meaningless in a moral sense, "academic" in the pejorative, to compose essays or study culture when that work is detached from consequences. Consequentiality might arise, though, from place-based or bioregional study as it couples with more personal modes of communication and the affective domain. Leaders of the environmental movement know this lesson well. Following their lessons, it becomes incumbent on us educators to activate ourselves, cultivate place-based attachments and sympathies, explore more personal modes of discourse in our own research and in our classes, and instill a renewed sense of significance to make the study of language more meaningful to students. Civic participation need not take radical forms, nor does it have to toe the line of political correctness along green parameters. Praxis in the language classroom only need restore the concept of action in that word's meaning. If compelled to adopt a denomination, I would favor the term "applied composition." In my own classes it might take the form of letters or editorials to regional, local, or national periodicals; petitions to leaders of industry, business or government; public performances of polemics or exhortatory discourse; appeals to members of environmental organizations; and excursions into participatory democracy of all sorts, beginning by examining the policies and practices of the college or university itself and extending from physical plant to language habits.

In our composition classrooms a skeptical regard for theory—which is as subject to planned obsolescence as our technology—might invite us to excavate cast-off practices. What, for instance, did we lose when we marginalized the personal narrative? And if belles lettres were ousted, as writing teachers turned from written product toward process and then toward post-process, maybe we could revive literary modeling and stylistic emulation. (By emulation, I mean that practice popular in previous centuries of imitating or emulating the prose of masters, whoever they are determined to be.) More to the point, any plan for restoring bioregions through composition will mean integrating more exigency-based discourse forms and more disparate disciplines. In a spirit of the interdisciplinary, I review the humanist tradition below, show how it erects barriers between our species and nature, draw upon some research from environmental educators, and examine the case of a researcher-activist who routinely incorporates critical and rhetorical tools in his work. In so doing I recommend ways my colleagues can restore a sense of consequentiality to their writing classes by encouraging a vital sense of place.

THE SHOW WE DROVE FROM TOWN

A sense of place, with an emphasis on natural space, has become all but lost to most of us today. If we gain it as a child, we often lose our sense of place in nature once we become adults and responsibilities begin to press upon us. We do not take the time, as a rule, to learn the features of our bioregions. In our work as scholars and teachers, we rarely study nature unless we teach the sciences. As humanists, we teach theoretical "humanities." Our built environments, more comfortable all the time thanks to technology, enclose and cushion us, embargo nature, and allow us to forget that we depend upon the natural world for food, water, air, and even for spiritual balance. Deborah Tall tells the striking story of a friend who, visiting his mother, got lost in the suburbs among streets indistinguishable in name and appearance: "all the houses an identical rosy beige, he drove around desperately in [his mother's] car, pressing the garage-door opener, waiting to see which door would open" (104–5). Few anecdotes capture the loss of place-connectedness as well. Our planetary population, now pushing eight billion, curbs meaningful opportunities to escape the bustle and hum that accompany civilization and urbanization. But more than populations and circumstances divide us from empirical wisdom of watersheds and ecotones, from biological economies and growing seasons, erosion rates and rainfall inches. Our separation from the waters and topographies of our homelands stems also from complex cultural shifts that tie to historical humanism and its dependence on technology.

Humanist assumptions underlie our family dynamics and consumer habits, tastes in arts and entertainments, our educations and the classes we teach. When R. F. Kennedy, Jr., addressed an auditorium full of people at my university, he ended his fine speech on "Our Environmental Destiny" with a set of humanist appeals: because so many American arts, belles lettres, and government tenets are based on nature, we should save natural spaces as a natural archive of our culture. As I use the term, *humanism* is a philosophical orientation or worldview that depends heavily on science, verifiable experience, and rationality; that valorizes philosophy, technology, and the arts; and that is generally suspicious of supernaturalism, religious ritual, and emotional exuberance. Kennedy was certainly right, but for the wrong reasons, or at least his rhetoric was ill chosen. He was appealing to his audience in humanistic ways, waxing anthropocentric, even "speciesist," as some environmentalists have begun to say.

More judgmental in his views of humanism, David Ehrenfeld defined it as "the religion of humanism: a supreme faith in human reason—its ability to confront and solve the many problems that humans face, its ability to rearrange both the world of Nature and the affairs of men and women so that human life will prosper" (5). With its emphasis on reason and rationality in particular, humanism has erected barriers between us and the natural world, and it has contributed to specializations within higher education, according to which the professionals with most legitimate claim to study nature are scientists, and thanks to which professors in English departments no longer speak a lingua franca, so disparate are the discourse communities. It has also, I allege, added to overspecialization in the academy. Richard Lanham has written, on style in our profession, "At its worst, academic prose makes us laugh by describing ordinary reality in extraordinary language" (4). The fascination with theory that often besets language studies is an unexamined outgrowth of humanism. What conversely attracts about ecocriticism is its usefulness as a methodology and its aversion to conform to theories, especially theory for theory's sake.

Overspecialization is a problem in American society at large and in the academic disciplines in particular. Alfred North Whitehead, writing in 1929, decried "the mediocrity of the learned world," which he traced to its "exclusive association of learning with book-learning" (Orr 1996, 19). Instead he endorsed "'first-hand knowledge,'" that is, knowledge gained through empirical experience rather than second-hand, thus discounting knowledge gained vicariously (19). Crippled by the vicariousness of our work, many of us language professors traffic in abstraction and conjecture. Rarely, though, are we called upon to demonstrate the verifiability of our concepts and claims. Hence we vex ourselves to develop rationalistic vehicles of assessment for the writing classroom in attempts to gauge the success of student writers. Such quantification is apt to be incompatible with composition, insofar as it cannot assess analogy making, passion, or spiritual growth. We render lip service to the interdisciplinary in the form of writing across the curriculum, but we implement it only rarely. Compartmentalized thinking in general is one source of our present environmental crisis. Science, likewise, far too often gets overlooked in the formation of public policy. Due to specialization, similarly, science is becoming increasingly politicized (Wilkinson; Ehrlich and Ehrlich), while some language theorists superficially adopt the language of the social sciences.

When intellectuals embraced humanism, many of them banished nature as a site of inspiration and wonder, a route to the truth. By and large it is not considered rational, and hence it is not humanistic, to regard contemplation of the natural world as a route to human improvement. Seventeenth-century Puritan poet and rhetorician John Milton, a canonical writer and pillar of Christian humanism, reveals an inherited attitude toward nature and its incompatibility with the human divine. In Milton's paradigm, nature (Satan in the snake's shape) occasioned Adam and Eve's tragic fall from grace. In his poetic hymn "On the Morning of Christ's Nativity," news of the newborn God-in-man drives pagan deities from the sacred groves, at the same time driving nature outside the reach of human reverence. Those suspect nature deities in Milton's devotional poem grow superfluous, shamefaced, tawdry, and frail beside humankind's resplendent savior-to-be, Jesus Christ. "Nature in awe to him / Had doffed her gaudy trim," which is to say that abject nature immediately deferred to Christ's greater qualifications and glory (ll. 32–33).

Significantly enough, nature throughout the poem is female, while Christ is male. She seeks "To hide her guilty front with innocent snow" (l. 38); that is, she mantles herself in December whiteness to feign innocence and purity. Similarly Shakespeare's King Lear, imagining a "simpering dame" in his mad interlude on the heath, raves about the "riotous appetite" she disguises—especially in that risky space below the waist, whose domain "is all the fiends'; there's hell, there's darkness, / There's the sulphurous pit, burning, scalding, / Stench, consumption" (4.6.114, 123–25). Nature is female, in the humanist equation. She is dissembling, sexually charged, deadly and dire. Vestiges of such attitudes remain today. Depraved men and savage beings hide in nature, women are sullied, and children drown or suffer attacks by beasts. Humans must be vigilant, wary of the wild, and zealous enough to effect a keen regeneration through violence when required. Humans reading such accounts may risk those same assumptions. Still today, reevaluating nature and culture may make a perilous journey for college students.

Milton's humanistic poem depicts nature, illuminated in the light of Christ, as the grotesque source of humanity's unfortunate fall from grace. *Genius loci* are involved in that fall. The origin of our species' sorrow, nature, is "Pollute[d] with sinful blame" and blasted with "foul deformities" (ll. 41, 44), for its complicity with humankind's lapse from grace. Correspondingly the sun, understood today as the daily progenitor of life on Earth, "hid his head for shame / As his inferior flame / The new-enlightened world no more should need. . ." (ll. 80-82). Here, Milton exercises great poetic license in depicting a sun overshadowed by the infant man-god. Here, Milton exemplifies the quintessence of humanism; our species does not need the sun, his reasoning proceeds, so great is humankind and the human god. Such theocentric fancy belies our ecocentric knowledge that death begets life. Spiritual life emanates from heaven, while "the old dragon under ground," that is, Satan, outgrowth of the planet's bowels, "wroth to see his kingdom fail, / Swings the scaly horror of his folded tail" (ll. 168, 171–72). That is to say, Satan and nature and the sun together conspire to create a kind of counter-kingdom to Heaven.

Planetary powers appear evil in Milton's worldview, and thus it falls on humans to overthrow that evil—through rationality, technology, and faith. The most telling instance of humanism's supremacy over nature occurs in Milton's poem with the revealing expulsion of animated nature when the Christ-child arrives. Then, from "haunted spring and dale," the "parting Genius is with sighing sent" (ll. 183, 185). A protector or guardian deity in classical antiquity, the "Genius" typically pertained to a particular place, as in the Latin *genius loci*, distinctive characters or atmospheres of a place with respect to the impression they make on the mind. Peculiarly enough, as if aware that Milton's intolerance of paganism flaws his art, the editors of *The Norton Anthology of English Literature* from 1993 furnished a quaint disclaimer. "With the coming of Christ all these picturesque local deities are dismissed, and the poet clearly regrets their departure" (Abrams et al. 1440, n9). "Regrets"? The editors furnished no evidence of the poet's regret, nor is that regret evident anywhere. Moreover, the editors' "picturesque" is irretrievably condescending. In a second apparent attempt to rescue Milton from biophobia and unregenerate humanism, the editors commented in a footnote, "Like ghosts at sunrise, the pagan gods, geniuses, and fays (fairies) are all bound to disappear at the rising of the Christian sun" (1442, n6). Neither Milton nor his editors deserve more blame than other writers of modern times, but they do furnish textbook examples of humanism's sway over indigenous place historically, and of our attendant alienation from nature today. More recent Norton editions omit those footnotes.

A sophisticated environmental education necessitates an ecological criticism, an ecocriticism. If scholars still are wary of trends from Europe, after some of the excesses of poststructuralism, ecocriticism nimbly has avoided those pitfalls so far. Contributors to *Interdisciplinary Studies in Literature and Environment*, a scholarly journal published since 1993, and the more recent *Journal of Ecocriticism*, are demonstrating the viability of this aversion. In the following bit of original ecocriticism, I draw attention, however briefly, to an obscure sonnet by Gerard Manley Hopkins. He is a favorite poet of green readers for the past one hundred years, even though his work celebrates nature in a way that redounds ultimately to the glory of God, and he is offered as a contrast to the pillar of humanism that is John Milton. Whereas Milton was a lay religionist, a Puritan, Hopkins worked as a Jesuit priest, and therefore my distinctions underscore the divisions between the Protestants and Catholics. Unconventionally, though, "The Sea and the Skylark," written in 1882, never makes mention of godhead and ends on a particularly evolutionary note that strives to reconcile Hopkins's brave strain of Christianity with the challenges of Darwinian evolution. Few critics have been willing to acknowledge the Darwinian strains in the poetry of Hopkins, but an ecocritical approach to this poem may allow contemporary readers to hear it anew.

The Sea and the Skylark

On ear and ear two noises to old to end
Trench—right, the tide that ramps against the shore;
With a flood or a fall, low lull-off or all roar,
Frequenting there while moon shall wear and wend.

Left hand, off land, I hear the lark ascend,
His rash-fresh re-winded new-skeinèd score
In crisps of curl off wild winch whirl, and pour
And pelt music, till none's to spill nor spend.

How these two shame this shallow and frail town!
How ring right out our sordid turbid time,
Being pure! We, life's pride and cared-for crown,

Have lost that cheer and charm of earth's past prime:
Our make and making break, are breaking, down
To man's last dust, drain fast towards man's first slime.

Hopkins contrasts an unnamed town with a skylark and the sea. The town, located some distance away from the setting and the speaker, is positioned as a human construct and a synecdoche for human civilization. The bird and sea, on opposite hands of the poem's speaker, function as synecdoches for nature—they stand for the whole of nature in its various guises. As the poem represents them, the town is static; the bird and the sea, dynamic and aesthetic. The tideline demarcates nature from civilization, separating the earthbound town from the elements of air and water, those respective media of the bird and the sea, a contrast that implies a perceptual conflict within the speaker of the poem.

The "music" of birdsong inspires the poet to simulate that music with his own medium, the medium of language, at the same time that it tests the limits of his facility with that language. Bird music is daring and innovative ("rash-fresh"); it is constantly renewing ("re-winded"); it is smooth and seamless as silk cloth ("new-skeinèd"). The bird-music is a portion of a "score." The bird writes on the sky in sound and motion alike; the bird's song accompanies the bird's rising and falling flight. Performer and conductor at once, the bird performs an aesthetic sleight-of-hand by stripping its cloth-music from a reel or wheel, that is, a "wild winch," a bobbin or a spool. Human tools and products of human tools, outgrowths of the town and civilization such as the winch, furnish ready reference points. Hopkins's keen attention to the Welsh "cynghanedd or consonant-chime" (265) might seem indulgent or excessive to modern ears. However, Hopkins considered the linguistic exuberance or delirium of the bard to be a close approximation of the bird.

Placing nature and civilization in conflict, as Hopkins does in this poem, suggests a criticism of the humanist project. I did some digging in etymologies and in other Hopkins poems, most famously his "God's Grandeur." Civilization, represented by the town in "The Sea and the Skylark," is shameful, shallow, frail, soiled, weary, smelly, materialistic, and even sick. Only in a bitterly ironic sense, then, is humankind the "crown" of creation. Not known for his irony, Hopkins may be read in a new light hereby. If the word "crown" denotes the pinnacle or the apex of achievement and glory, it also calls up the image of the crown of thorns Christ wore in passion when he shouldered the sins of humankind. We—humans and our civilization—"crown" nature only with our own ignominy. Nature in the poem, as embodied by the bird and by the ocean, contrasts with civilization, clashes with it, inasmuch as nature is charming, spontaneous, cheerful, powerful, fresh, unrestrained, musical, artless, and pure. The bird and the sea are beings, genius loci, vibrant and alive. By contrast, the town and the human efforts it represents (by means of a clever synecdoche) are dead. They are moribund in ways that Christian conceptions of a fall from grace in the Garden of Eden fail adequately to explain.

Dense and compelling, "The Sea and the Skylark" swells richly with conflicting sorrows and joys. It does not rely on Christian readings of its imagery. Its intellectual author could synthesize his Christianity and the challenges of Darwinian evolution. Nowhere in the Hopkins poem is human fallibility ("this shallow and frail town") attributed to Adam or Eve, those weak beings who yielded to the rebel angel Satan—quite the opposite. The poem's great surprise lies in the final phrase, "first slime," an inescapable allusion to the then-recent scientific theory of Darwin, whereby the miraculous cycle of life that gave rise to humankind began as sea-slime. Notice how *slime* cooperates with *trench* and *drain*: lower life forms puddle and ooze back to the primordial soup. Our species and civilization, unworthy and lacking imagination, devolve back to the slime that gave rise to us. That same slime is draining to the sea where it (and we) may start over again at the place we began.

To acknowledge this 137-year-old poem by a Jesuit priest as environmental literature requires the reader to read it differently. Reading it differently calls into play some innovative critical approaches, some eco-critical skills. Interrogating the legacy of humanism is one way to sharpen those skills. Literature of the environment is coming to be redefined in innovative terms, both for the devices by which it reproves business and industry and for the ways it historically has venerated the rights of nature. Scholars of Hopkins slowly are coming to recognize the ways he had begun to reconcile evolutionary science. (See, e.g., the Cary H. Plotkin article in the Gerard Manley Hopkins Archive.)

Without directly attributing such alienation to humanism, many other nature writers and environmentalists have lamented what we lost when we tossed our erstwhile reverence for natural places. If college teachers were to integrate environmental literature into their classrooms, a sense of place could be enhanced. Annie Dillard, the 1974 Pulitzer Prize-winner for nonfiction, deplores our species' resolute silencing of nature. Writing as a Christian herself, she indicts the Judeo-Christian tradition for teaching us a lesson we have taken too hard to heart. That lesson is "the muteness of the human stance in relation to all that is not human" (92). For most witnesses except a select few granolas and scientists (odd bedfellows if ever there were), such muteness in the face of nature seems routine. In her gentle jeremiad Dillard sadly notes, "It is difficult to undo our own damage, and to recall to our presence that which we have asked to leave. It is hard to desecrate a grove and change your mind" (88). The damage, culturally speaking, has been done. Still, aiming college courses toward the job of restoring those desecrated groves is something we may undertake; it is a challenge worthy of the high stakes. Observe closely, Dillard recommends. Attend as if attention were an epistemology, for until the show we drove from town returns, "until the pagan gods slip back to their hilltop groves, all we can do with the whole inhuman array is watch it" (90). Watching so closely would involve a mystical process that is alien to theoretical humanists; it requires "sacrifice, the suppression of self-consciousness, and a certain precise tilt of the will, so that the will becomes transparent and hollow, a channel for the work" (88). Most of us find it hard to imagine college curricula straying very far from their comfort zones unless they were to embrace the emerging eco-consciousness, unless they effect a merger with environmental education. Corroborating Dillard, Deborah Tall contends, "Judaism is the religion that by and large defused the religion of sacred place" (106). Tall tackles this tender subject with the insight of one raised in the Jewish faith. "The pagan gods the early Jews set out to overthrow were the numerous place-defined, local nature deities," she writes, as we see so abundantly in Milton and in other writers who shaped Judeo-Christian humanism.

REGAINING A SENSE OF PLACE

That future might appear increasingly global and mobile, but this appearance need not dissuade language teachers from urging students to restore a sense of place. David Orr asserted that we may remedy much contemporary angst and disaffection by helping to develop "a deep concept of place as a repository of meaning, history, livelihood, healing, recreation, and sacred memory and as a source of materials, energy, food, and collective action" (1994, 163). Failure to do so will perpetuate the trend toward becoming not merely "'dis-placed' people who are physically removed from their homes," but indeed "'de-placed' people, mental refugees, homeless wherever we are" (163)—a perception that is worth exploring for the possible benefit of the criminals, homeless, and mentally ill in America today. Here, again, Deborah Tall's chilling story of the garage-door opener comes to mind. Lost in a human-centered world, we lose our topographic bearings and some of ourselves as well. Teachers of literature and rhetoric can begin to establish a place-based awareness among students by assigning them to read writers who explored deeply the *genii loci* of beloved places and sometimes worked to preserve them—Sarah Orne Jewett (on late-nineteenth-century Maine), John Muir (on Yosemite and the Sierra-Nevada), Sigurd Olson (on Minnesota, Wisconsin, and the Great Lakes), Edward Abbey (on the desert Southwest), and Linda Hasselstrom (on South Dakota). Such writers can model the healthy desire to affiliate oneself with a region—named or unnamed, cultivated or wild, small or large, lush or dry. To affiliate is "to bring or to receive into close connection as a member or a branch," "to connect or associate oneself." The root word is Latin *filius*, meaning *son*, which suggests (when applied to place relations) actual genetic connections to a district of desire. Related emotionally if not etymologically are the Greek suffixes *-philia*, meaning friendship, and *-philos*, translated as relatedness. This Greek suffix has picked up some unfortunate connotations of pathology, as in *necrophilia* or *pedophilia*, thereby partially eclipsing the more healthy denotations of intimacy and attraction that inform *biophilia* and *philosophy*. For the ends of ecocriticism or applied writing, these Latin derivations demonstrate that what we affiliate with is yoked to where we've lived.

As is well known in ecocritical circles, Harvard biologist E. O. Wilson in 1984 rejuvenated the roots of the word *biophilia,* which appears in too few standard dictionaries but first was used by psychoanalyst Erich Fromm. Wilson's biophilia hypothesis holds that "'the connections that humans unconsciously seek with the rest of life' are innate" (Orr 1994, 46). To whatever degree life remains tied to the land—through air, water, animals or plants for food—biophilia can broadly suggest a connectedness and an affiliation with specific places as well. *Topophilia,* a related term from geographer Yi-Fu Tuan, to denote "all of the human being's affective ties with the material environment" (10), would be nice to engender in students, if it is not too late for most of them to remember place-love in some collective unconscious. As Orr lamented, however, reflecting upon education today, "Locality has no standing in the modern curriculum" (1994, 129). In literary studies, for instance, the closest focal point is sectionalism or regionalism—used somewhat pejoratively to mean local color—a subspecies of literature typified by vernacular speech patterns and slang, quaint folkways and folklore, and rural characters, rustics, bumpkins. Scholars and critics still rarely condescend to treat local-color literature and characters as anything much more than antiques or curios. Decrying the lack of serious locality studies in higher education today, Orr created a neologism, *biophobia,* which he defined as "the culturally acquired urge to affiliate with technology, human artifacts, and solely with human interests regarding the natural world" (131). Anyone who has witnessed the fascination of people with smartphones today will understand fully.

If biophobia is antithetical to the purposes of environmental education, biophilia is essential—not that students can be taught to love life, only given the opportunity and shown the ropes and rewards through the study of place. While environmental education is a relatively new field, most practitioners seem to agree sadly with Orr: "If by some fairly young age . . . nature has not been experienced as a friendly place of adventure and excitement, biophilia will not take hold as it might have. An opportunity will have passed, and thereafter the mind will lack some critical dimension of perception and imagination" (143). "Perception and imagination," perhaps the most pertinent words in this passage, lead to the practice of using place as the basis for prompting critical thinking in courses. Adult learners can be taught to bring their perceptions and imaginations to bear on the place-based materials of history, religion, public policy, psychology, and personal memory in ways that younger students perhaps cannot. If youthful opportunities often pass too fast to establish the strongest bioregional affiliations, the greater range of older students' experiences and interests can serve nonetheless as a forceful surrogate. The more sophisticated passions and ambitions of adult learners have their own compensations, and they might lend themselves more fully to restoring bioregions through applied study.

The desire to affiliate, whether intellectually or emotionally, can reverberate with power for students of the language arts. The interests of environmental educators and ecocritics can come together. Mitchell Thomashow, for his provocative study *Ecological Identity*, drew from related forms of identity politics and argued that developing a profound sense of oneself in relation to natural ecosystems supplies a necessary foundation for the labor of environmental advocacy. Thomashow calls this development of oneself "ecological identity work" (5). Adapting psychological and spiritual models, Thomashow held that "ecological identity describes how we extend our sense of self in relation to nature, and that the degree of and objects of identification must be resolved individually" (3). Analyzing oneself as an outgrowth of place becomes a challenge to be met by studying—almost inevitably through forms of writing—the particulars of one's ecological origins. Just as "family of origin" studies have proven fertile approaches to counseling psychology, Thomashow's ecological identity practice shows a twin promise, both for maturing in adult learners those affiliations that have been shortchanged in youth, and for helping to vanquish youthful disaffiliations with family and place. Ecological identity work also could help to restore some of the lost consequentiality of language classes. The tantalizing possibility exists for teachers of rhetoric and literature that ecological identity work may reintegrate an estranged citizenry and help restore degraded ecosystems if such writing becomes a habit.

Underlying Thomashow's compelling study is the belief that "It is the personal introspection that drives one's commitment to environmentalism" (5). Far from the Protestant model of self-examination, as practiced by the American Puritans especially, ecological identity work does not seek to dwell on sins and commitments, lapses and leaps of faith, but rather on the social and biophysical particularities of one's place of origin. William Wordsworth, himself a powerful advocate for the particularities of nature in the Lake District of eighteenth- and nineteenth-century England, recommended drawing upon "emotions recollected in tranquility" to recapture the often-fleeting and emotional fluxes of childhood (148). Wordsworth's poetry is vividly place-based. Yet recollecting emotions in tranquility today not only is not part of most current college curricula, it is tacitly numbed and nullified by the pace of cinematic entertainments, the blitz of television imagery to which we all fall susceptible, and by incessant print and audio persuasions to partake in pop-culture trends and fads. Place-based reflection is rare.

Thomashow finds adult learners very interested, when asked to recount their most memorable environmental experiences, in exploring "childhood memories of special places, perceptions of disturbed places, and contemplations of wild places" (7). This narrow range of recollected emotions, all locked into perceptions of place, implies a dialectic or process to be worked through before synthesis can be attained. If memories of special places and wild places are devoutly to be wished for and affiliated with, in our curricula as it is shaped today, then disturbed places naturally will be regretted and reviled. The eye despises what repels the heart and mind. The urgent introspection of ecological identity work, akin to the solitude that nature writers often undergo, "is not framed as alienation and atomism but refigured as affinity and integration" (Roorda 404). The upshot in writing and literature students might be imaginative projection toward a more restorative time, or a new or renewed commitment to community service or activism.

APPLYING COMPOSITION ON THE GROUND

Efforts to restore damaged bioregions through applied composition need not reinvent the proverbial wheel. A growing body of research in integrative studies, and in rhetoric and literature, analyzes the ways that teachers may legitimately incorporate service learning in their courses. (Defined variously as paid or unpaid action taken to enhance the community, service learning typically complements one's major field.) If teachers combined service learning with curricular opportunities to allow students to affiliate with nearby bioregions, indeed if ecocriticism and ecocomposition courses were designed dynamically to acquaint students with local issues and to build on affiliations, then the emotional growth of those students could potentially prove a powerful stimulus for their writing and imbue classes with that missing consequentiality. Terrell Dixon and Lee Smith, working with disaffiliated students in downtown Houston, Texas, have explored some forms of service learning that they allege have reactivated composition:

> Students work together in teams to read, analyze, and write essays on urban environments and ecology. In addition, they work with environmental groups and agencies in the community, such as the Galveston Bay Foundation, an inner-city park, an arboretum, and the Local Emergency Planning Commission, and write documents specifically for these groups, like reports, trail guides, press releases, and educational brochures. Students also keep journals in which they reflect on their service learning experiences. (A3)

In the model Dixon and Smith practice, students begin by augmenting their analytical and collaborative skills; next students discover organizations that can benefit from those skills; and then they apply those skills on behalf of the organization, all while they are exercising a common form of metacognition—writing a reflective journal—that appears to be a first step toward Thomashow's more ambitious and more deeply spiritual ecological-identity work. Dixon and Smith refer to their experiment as an "innovative blend of service learning and ecocomposition with an urban focus," one that proves challenging and rewarding because it "immerses students in a public discourse with its particular rhetorical demands, preparing them more effectively to write in the workplace" (A3). This model, then, is one that aims pragmatically toward the demands of professional communication. Working collaboratively, the Houston students learn to develop communities at the same time they develop their cooperative skills.

Erik Ryberg embodies this concept of applied composition. A lawyer now by professional bent, an environmental activist when I met him, he studied for two years toward his PhD in English language and literature at the University of Virginia before dropping out to try his hand at saving the planet. He spent some months in jail for civil disobedience. The native of Oregon worked full time to restore at-risk ecosystems and bioregions in the abundant public lands of the West. He tramped across those lands, bearing witness firsthand to the damage inflicted by timber-industry clearcuts and Forest Service logging roads, private ranchers grazing cattle, and off-road vehicle enthusiasts pursuing combustion-fueled sports. To expose that damage, he published an illustrated and well-written newsletter entitled *The Stump*, a punning title that squints down two paths—toward the ugly remnants of industrial logging, and toward that place from which declamatory notices issue. Interested especially in forest ecology, Ryberg worked for several "ad hoc citizens advocate groups" and as a researcher in Montana, Idaho, and Washington for the Missoula-based Ecology Center. He took time off from a demanding schedule in law school and agreed to a telephone interview. In particular I was curious to learn from him how he believes his work connects with place, is inspired by place. Also I wanted to learn from him how, as a professor and a Sierra Club leader, I could use the tools of ecocriticism to entice my students and volunteers to become front-line researchers and activists like him. My higher purpose would be to encourage spiritual connections to place and translate those connections into ecological consequentiality.

On the matter of place, he chastened me for believing that I could ever presume to choose a place to which my students truly would be able to attach. Ryberg is pessimistic about the prospect of field trips instilling place-based affiliations. He is certain that

> those sorts of things never worked for me. Having somebody take me somewhere never really worked. It always ended up being their place. I got most involved with a place when chance arose from doing my own research. One of the best ways to get somebody fired up about a place, and this trick was pulled on me a long time ago, is to find people who are already interested in forest conservation and give them an EA [environmental assessment] to read. Then let them know that if they want to do anything about the timber sale that's about to destroy that place, they should consider commenting on the projects and writing an appeal because *nobody else is going to do it* (italics mine).

When Ryberg uses words like those I have placed in italics above, there is a sense of consequentiality. How many of us English educators can say the same for the latest article, textbook, or theory we are advancing—that no one else is going to do it? (I remember as a graduate student abandoning Faulkner as a dissertation topic when I discovered a published book entitled *Research Opportunities in Faulkner Studies*.) And even if we can say with all honesty that nobody else is going to take up the work that we are undertaking, who among us could say that that work would be tangible, consequential, or significant in any fashion comparable to preserving a forest ecosystem?

The work Ryberg so often undertook, the work of scrutinizing agency-written documents known as "environmental assessments" or "impact statements," has immediate significance. It furnishes bases on which to write appeals, and writing appeals saves forests. Ryberg concluded on place by observing, "there's a kind of responsibility that you have, and you also have the power if you choose to exercise it." He was talking about responsibility, democracy and citizenship, the power of artful writing and speech, to effect change in the face of sciences that are compromised and politicized. English teachers bear a related responsibility to teach consequential literature and writing well.

The process by which Ryberg reads, writes appeals, and halts timber sales and grazing allotments warrants attention in the context of composition, but first I would give some background on the relevant political processes. Agencies like the Bureau of Land Management and the Forest Service justify themselves by selling products (e.g., grass for grazing, pine and fir trees for timber) to ranchers, loggers, and other resource-dependent interests. Because the lands those agencies administer are public, held in the public trust, the products yielded by those lands are sold at subsidized rates. The hitch is that the chief beneficiaries are corporations or at least big operators. Until recently no one cared if a rancher, for example, grazed a cow-calf pair on Federal lands for $1.35/month—or even if they refused to pay their grazing fees for decades, which lies at the core of the Cliven Bundy debacle that unfolded in Nevada in 2014. As environmental awareness has risen in recent decades, however, so have competing claims for the best use of public lands. Campers as a rule do not like to sleep among cow-pies, nor do fishing enthusiasts care to see the banks of favorite trout streams eroded by Hereford hooves, the water muddied by them. Activists like Ryberg, using language effectively, can "aim" their love for wild places toward grazing allotments and timber sales, which they may challenge in writing.

Analytical and evaluative skills developed by studying rhetoric, literature, technical writing, or literary criticism come in handy, as one might imagine, for reading between the proverbial lines of bureaucratic documents that offer lush acres of our native birthright for sale or lease. Writer-researchers like Ryberg who love the land can wield their skills to restore degraded bioregions or prevent the initial degradation. They engage through firsthand knowledge with the elusive real world. Working for modest salaries or wages, they make service a livelihood; they enact the principles of democracy and practice citizenship in ways we hope that service learning will. "We call it activism, but it's really just part of being a citizen; it shouldn't have such a grand name." By any other name, "activism is a very fertile ground" for applying writing skills. And "it is probably the duty of a rhetoric professor," he lectured me, "to show examples of what to do with those skills if you want to." If ecocomposition and ecocriticism list toward professional communication (at least as we see it in the work of Dixon and Smith), it can justifiably list as well toward enrichment in the liberal arts. Students, Ryberg notes, "shouldn't be made to learn critical thinking and writing skills just to use them in the service of Microsoft or other corporations. Students should be taught that they can use them however they want." And "however they want" I take to mean that students should not have to toe the line of political correctness along green parameters, nor should they have to practice critical thinking necessarily in the service of the writing teacher's political inclinations.

How does Ryberg apply his writing skills? "There's a lot of writing associated with my job, and so it helps to be able to write clearly." He composes "appeals, legal briefs when we sue, information requests, propaganda pieces, letters to the editor, columns, and so forth." His candor is refreshing. Specifically I was interested in learning how he goes about the appeals process. He looks at government documents, he says, for "tensions, and then you try to find or try to suppose what those tensions are about, what might be going on behind the lines and behind the text." Such presuppositions about agency dishonesty arise from his belief that "a bureaucracy like the Forest Service uses language to make reality as irrelevant as it can." Language for agency writers, Ryberg believes, is intended "to cover up or shade over what's actually happening on the landscape. And the job of an activist is to plow through that language and find out what's really happening." Here Ryberg sounds like a poststructuralist literary critic, an inquisitive reader who applies a hermeneutics of misgiving to every text he reads. The difference is that such doubtful regard is arguably more warranted, more justifiable, in the case of documents whose writers have so very much to gain by enacting subterfuge. Literary artists are "creating something that provides an immense amount of joy, while writers of EAs are doing it wholly for themselves." Sounding like a disciple of Jacques Derrida, Ryberg recommends that "you have to take what they're telling you, and you have to find where the tension is and where the absences are. You have to learn to read the text as if it's concealing something." Such telltale signs are difficult to discern and more difficult to describe, but they are "the equivalent of watching body language when someone is talking." Unlike many a practicing literary critic or composition theorist, Ryberg is performing a discernible service to his communities and to future generations. His work is consequential, significant, and more rewarding for all of that.

CONCLUSION

To gain greater consequentiality, the principles of rhetoric and composition need to be applied. Students yearn to discern results in their work, to see outcomes beyond well-crafted sentences and convincing persuasive discourse, payoffs that can be assigned no precise value in the marketplace. Trends in literary, composition, and language studies of recent decades—away from symbols, themes, and morals and toward race, class, and gender—constitute gestures in the right direction. But they remain chiefly gestures due to an alienating reliance on theory and abstraction, overly specialized language, ingenuity, and invention for invention's sake. Symbolic and largely insignificant, such trends remain humanistic, anthropocentric, even speciesist. The emerging ecocomposition and ecocriticism, with a greater emphasis on consequences, will demand that readers and teachers selectively integrate principles of environmental education, a field of study that is gaining maturity and sophistication as crises like global climate change come into the popular consciousness and into policy debates. Such sea changes likewise will require us to expand our spheres of sympathy to include other species and bioregions, and to extend our ethical regard to nature and nature's rights. There is a hopeful future for all life on this planet with these new forms of knowledge, and ecocritics can help to shape it.

Acknowledgments

All of these writings have undergone extensive revision and supplementation since their first appearance, and all of them have benefited from the attention of the original editors.

"From Sublimity to Ecopornography: Assessing the Bureau of Reclamation Art Collection," *Journal of Ecocriticism*, 2009.
"Gifts and Misgivings in Place," in *On Location: Adventures in Place-Based Teaching* (Reno: University of Nevada Press, 2008).
"An Iconography of American Sabotage," in *Nature et Progrès: Interactions, Exclusions et Mutations* (Paris: Presses de Universitaires, 2006).
"Language and Landscape in Washington Territory," *American Transcendental Quarterly*, 2004.
"Greening the Dramatic Canon: Henrik Ibsen's *An Enemy of the People*," *Interdisciplinary Literary Studies*, 2001.
"Restoring Bioregions through Applied Composition," in *Ecocomposition: Theoretical and Pedagogical Approaches* (Albany: State University of New York Press, 2001).
"Considering the Canon: American Nature Writing and the Wise-Use Movement," in *Western Futures: Perspectives on the Humanities at the Millennium* (Reno: University of Nevada Press, 2000).
"Rage Against the Machine: Edward Abbey and Neo-Luddite Thought," in *Coyote in the Maze: Tracking Edward Abbey in a World of Words* (Salt Lake City: University of Utah Press, 1998).
"Literary Activism and the Bioregional Agenda," 1996; rpt. in *The First Decade of Ecocriticism from* ISLE*: Charting the Edges* (Athens: University of Georgia Press, 2003).

Works Cited

Abbey, Edward. 1979. *Abbey's Road*. New York: Plume, 1991. Print.
———. *Beyond the Wall: Essays from the Outside*. New York: Holt, Rinehart and Winston, 1984. Print
———. *Confessions of a Barbarian: Selections from the Journals of Edward Abbey, 1951-1989*. Ed. Dave Peterson. Boston: Little Brown, 1994. Print.
———. *The Fool's Progress*. New York: Henry Holt, 1988. Print.
———. "Forward!" *Ecodefense: A Field Guide to Monkeywrenching*. Ed. Dave Foreman and Bill Heywood. Tucson: Ned Ludd, 1987. ii-xi. Print.
———. *Good News*. New York: Dutton, 1980. Print
———. *Hayduke Lives!* Boston: Little Brown, 1990. Print.
———. *The Journey Home: Some Words in Defense of the American West*. New York: Dutton, 1977. Print.
———. *The Monkey Wrench Gang*. New York: Lippincott, 1975. Print.
———. *One Life at a Time, Please*. New York: Henry Holt, 1988. Print.
Abrams, M. H., et al., eds. *The Norton Anthology of English Literature*. 6th ed. Vol. 1. New York: Norton, 1993. Print.
Achenbach, Joel. "Publish or Perish: The Untold Story of Thomas Harriot, the Greatest Scientist You Never Heard of." NationalGeographic.com. May 2003. Web. April 15, 2014.
Adams, Carol J. *The Pornography of Meat*. New York: Continuum, 2003. Print.
Adelman, Jeremy. "Early Doukhobor Experience on the Canadian Prairies." *Doukhobor Genealogy Website*. 2014. Web. May 3, 2014.
Adler, Thomas P. "Conscience and Community in *An Enemy of the People* and *The Crucible*." *The Cambridge Companion to Arthur Miller*. Ed. Christopher Bigsby. Cambridge: Cambridge University Press, 1997. 86-100. Print.
The American Artist and Water Reclamation. Washington, D.C.: United States Government Printing Office, 1973. Print.
———. Bureau of Reclamation, Fine Arts Collection. June 30, 2011. Web. June 4, 2014.
American Civil Liberties Union Blog of Rights. "H.R. 347." August 29, 2013. Web. June 11, 2014.
Androsoff, Ashleigh Brienne. "Spirit Wrestling: Identity Conflict and the Canadian 'Doukhobor Problem,' 1899-1999." Diss., Department of History, Univ. of Toronto, 2011.
Arnold, Ron. *At the Eye of the Storm: James Watt and the Environmentalists*. Chicago: Regnery Gateway, 1982. Print.
Arnold, Ron, and Alan Gottlieb. *Trashing the Economy: How Runaway Environmentalism is Wrecking America*. Bellevue, WA: Free Enterprise, 1993. Print.
Austin, Mary. *The Land of Little Rain*. New York: Dover, 1996. Print.
Baden, John A. "The True-Mann's West: Endangered and Forsaken?" *The Next West: Public Lands, Community, and Economy in the American West*. Covelo, CA: Island, 1997. 107-29. Print.
Barcott, Bruce. "The Elwha's Last Dam Summer." *Seattle Met*. July 22, 2011. Web. May 21, 2014.
Barnhill, David. "Taxonomies versus Ecosystem Approach." University of Wisconsin Oshkosh. Aug. 14, 2010. Web. April 22, 2014.
Barrow, Mark V., Jr. *Nature's Ghosts: Confronting Extinction from the Age of Jefferson to the Age of Ecology*. Chicago: University of Chicago Press, 2009. Print.

Bartram, William. *Travels of William Bartram*. 1791. Ed. Mark Van Doren. New York: Dover, 1955. Print.
Bass, Rick. *The Ninemile Wolves*. New York: Ballantine, 1992. Print.
———. "20515 House, 20510 Senate." *The American Nature Writing Newsletter* 5 (Spring 1993): 3-5. Print.
Bate, Jonathan. *Romantic Ecology: Wordsworth and the Environmental Tradition*. London: Routledge, 1991. Print.
Bergthaller, Hannes. "Minting New Ore." Unpublished essay, 1998. Print.
Berrian, Allegra. "U.S. Show at 2 Colleges." *The Spokesman-Review* [Spokane, WA] Sept. 15, 1974: 16-17. Print.
Berry, Wendell. *A Continuous Harmony: Essays Cultural and Agricultural*. New York: Harvest/Harcourt, 1972. Print.
Berryman, John. *The Dream Songs*. New York: Farrar, Straus and Giroux, 1969. Print.
Best, Steven, and Anthony J. Nocella, eds. *Terrorists or Freedom Fighters? Reflections on the Liberation of Animals*. New York: Lantern, 2004. Print.
Bishop, James Jr. *Epitaph for a Desert Anarchist: The Life and Legacy of Edward Abbey*. New York: Athenaeum, 1994. Print.
Black Elk and John G. Neihardt. *Black Elk Speaks: Being the Life Story of a Holy Man of the Oglala Sioux*. Lincoln: University of Nebraska Press, 1932. Print.
Blumenfeld, Y. "Protection of the Countryside." *Editorial Research Reports 1971*. Vol. II. Washington: CQ Press, 1971. Retrieved July 20, 2007, from CQ Electronic Library, CQ Researcher Online. Dead link. June 5, 2014.
Bly, Robert. *American Poetry: Wildness and Domesticity*. New York: Harper & Row, 1990. Print.
———. *Iron John: A Book About Men*. Reading, Massachusetts: Addison-Wesley, 1990. Print.
Boag, Peter. "Thomas Moran and Western Landscapes: An Inquiry into an Artist's Environmental Values." *Pacific Historical Review* 67.1 (Feb. 1998): 40-66. Print.
Boyle, T. Coraghessan. *The Road to Wellville*. New York: Penguin, 1994. Print.
Brazile, Donna. "Cliven Bundy, the Wild West hero. Or not." CNN Opinion. April 26, 2014. Web. April 28, 2014.
Brown, Robert F. "Regional Report: New England." *Archives of American Art Journal* 27.3 (1987): 38-9. Print.
Bryan, Frank, and John McClaughry. *The Vermont Papers: Recreating Democracy on a Human Scale*. Chelsea, VT: Chelsea Green, 1989. Print.
Budiansky, Stephen. *Nature's Keepers: The New Science of Nature Management*. New York: Free, 1995. Print.
Buell, Frederick. *From Apocalypse to Way of Life: Environmental Crisis in the American Century*. New York: Routledge, 2003. Print.
Buell, Lawrence. *The Environmental Imagination: Thoreau, Nature Writing, and the Formation of American Culture*. Cambridge: Harvard University Press, 1995. Print.
Buhle, Paul, and Edmund B. Sullivan. *Images of American Radicalism* 2nd ed. Boston: Christopher, 1999. Print.
Byron, George Gordon. "Song for the Luddites." *Orion*. Loyola U. Chicago. Aug. 2012. Web. May 7, 2014.
Cadillac Desert: Water and the Transformation of Nature. Dir. Jon Else. Prod. KTEH/San Jose and Trans Pacific Television. Alexandria, VA: PBS Video, 1997.
Calahan, James M. "Sabotage vs. Terrorism: Was Ed Abbey an "ecoterrorist?" Abbey-Web. Sept. 18, 2013. Web.
Callahan, Tim. "Environmentalists Cause Malaria! (And Other Myths of the 'Wise Use' Movement)." *The Humanist: The Magazine of Critical Inquiry and Social Concern* 55 (Jan./Feb. 1995): 10-15. Print.
Callicott, J. Baird, and Robert Frodeman. "Ecotage and Ecoterrorism." *Encyclopedia of Environmental Ethics and Philosophy*. Vol. 1. Detroit: Macmillan, 2009. 286-91. Gale Virtual Reference Library. Web. May 17, 2014.

Cantrill, James G., and Christine L. Oravec, eds. *The Symbolic Earth: Discourse and Our Creation of the Environment*. Lexington: University of Kentucky Press, 1996. Print.
Callenbach, Ernest. *Ecotopia: The Notebooks and Reports of William Weston*. 1975. Rpt. New York: Bantam, 1990. Print.
Cantwell, Robert. *The Hidden Northwest*. Philadelphia: J.B. Lippincott, 1972. Print.
Carson, Rachel. *Silent Spring*. Boston: Houghton Mifflin, 1962. Print.
Civil Action, A. Dir. Steven Zaillian. Perf. John Travolta and Robert Duvall. Touchstone, 1999.
Colby, Elbridge. *Theodore Winthrop*. New York: Twayne, 1965. Print.
Condit, Carl. *American Building Art: The Twentieth Century*. New York: Oxford University Press, 1961. Print.
Corcoran, James. *Bitter Harvest: The Birth of Paramilitary Terrorism in the Heartland*. New York: Penguin, 1990. Print.
"Countering Eco-Terrorism in the United States: The Case of 'Operation Backfire'." Final Report to the Science & Technology Directorate, U.S. Department of Homeland Security. Sep. 2012. Web. May 6, 2014.
Crosby, Alfred. *Ecological Imperialism: The Biological Expansion of Europe, 900-1900*. Cambridge, Eng.: Cambridge University Press, 1989. Print.
Cronon, William, ed. *Uncommon Ground: Rethinking the Human Place in Nature*. New York: Norton, 1995. Print.
———. "Introduction: In Search of Nature." In William Cronon, ed. *Uncommon Ground*, 23-56. Print.
———. "Telling Tales on Canvas: Landscapes of Frontier Change." In Jules David Prown, ed. *Discovered Lands, Invented Pasts*, 37-87. Print.
Curtis, George William. "Biographical Sketch of the Author." Preface to Theodore Winthrop, *Cecil Dreeme*. Boston: Ticknor and Fields, 1861. 5-19. Print.
DamNation. Dir. Ben Knight and Travis Rummel. With Ben Knight, Travis Rummel, Mikal Jakubal, Bruce Babbit, Kevin Yancy, David James Duncan, Elmer Crow, Jim Waddell, and Katie Lee. Felt Soul Media, 2014. Film.
Davis, Claire. *Winter Range*. New York: Picador, 2000. Print.
Deal, Carl. *The Greenpeace Guide to Anti-Environmental Organizations*. Berkeley: Odonian, 1993. Print.
deArmond, Paul. Deceased. E-mail correspondence 11 April 1996.
Delicath, John W., and Kevin Michael Deluca. "Image Events, The Public Sphere, and Argumentative Practice: The Case of Radical Environmental Groups." *Argumentation* 17 (2003): 315-333. Print.
Demers, David. "Semantic Noise." *Dictionary of Mass Communications and Media Research*. Spokane, WA: Marquette, 2005. 216. Print.
Devine, Robert. "Environment: The Cheatgrass Problem." *The Atlantic* 271 (May 1993): 40, 44, 46-8. Print.
DeWitt, John. "Reclamation and the Creative Artist." *Reclamation Era* 59.2 (1973): 12-15. Print.
Diamond, Jared M. *Guns, Germs, and Steel: The Fates of Human Societies*. New York: Norton, 1997. Print.
———. *Collapse: How Societies Choose to Fail or Succeed*. New York: Viking, 2005. Print.
Dillard, Annie. "Teaching a Stone to Talk." *Teaching a Stone to Talk: Expeditions and Encounters*. New York: Harper Perennial, 1982. Print.
———. *Pilgrim at Tinker Creek*. New York: Harper, 1974. Print.
DiPetto, Lisa. "Joe Hill." *Wobblies! A Graphic History of the International Workers of the World*. Ed. Paul Buhle and Nicole Schulman. New York: Verso, 2005. 131-42. Print.
Dixon, Terrell, and Lee Smith. "Service Learning and Environmental Writing." *MLA Newsletter* 30.2 (Summer 1998): A3-A4. Print.
Dominy, Floyd. *Lake Powell: Jewel of the Colorado*. Washington, D.C.: U.S. Department of the Interior, 1965. Print.
Dorson, Richard M. *Jonathan Draws the Longbow*. 1946. Rpt. New York: Russell & Russell, 1970. Print.

———. *America Begins*. New York: Pantheon, 1950. Print.

"Doukhobors." *The Canadian Encyclopedia*. Dec. 17, 2013. Web. May 3, 2014.

Dubofsky, Melvyn. *We Shall Be All: A History of the Industrial Workers of the World*. Urbana: University of Illinois Press, 2000. Print.

Dugatkin, Lee Alan. *Mr. Jefferson and the Giant Moose: Natural History in Early America*. Chicago: University of Chicago Press, 2009. Print.

Durfee, Jessica. "Images of Nature in Organizational Stakeholder Publications: The Rhetorical Function of Ecopornography in Greenwashing." Communication Association Conference, New Orleans, LA, November 23, 2002. Web (dead link).

Earth First! "Wolf Hunt Sabotage Manual." Oct. 4, 2013. Web. May 14, 2014.

Easterbrook, Gregg. *A Moment on the Earth: The Coming Age of Environmental Optimism*. New York: Viking Penguin, 1995. Print.

Edward Abbey: A Voice in the Wilderness. Dir. Eric Temple. Temple Productions, 1993.

Edwards, J. Gordon. "Malaria: The Killer That Could Have Been Conquered." *21st Century Science and Technology* 6 (1993): 21-35. Print.

Ehrenfeld, David. Foreword. *Greening the College Curriculum: A Guide to Environmental Teaching in the Liberal Arts*. Ed. Jonathan Collett and Stephen Karakashian. Washington, DC: Island, 1996. ix-xi. Print.

———. *The Arrogance of Humanism*. New York: Oxford University Press, 1978. Print.

Ehrlich, Paul R. *The Population Bomb*. San Francisco: Sierra Club, 1969. Print.

Ehrlich, Paul R., and Anne H. Ehrlich. *Betrayal of Science and Reason. How Anti-Environmental Rhetoric Threatens Our Future*. Washington, DC: Island, 1996. Print.

Elder, John. *Imagining the Earth: Poetry and the Vision of Nature*. Urbana: University of Illinois Press, 1985. Print.

Eliot, Jr., Ellsworth. *Theodore Winthrop*. New Haven: Yale University Library, 1938. Print.

"Elizabethan Bear and Bull Baiting." ElizabethanEra.org. 2012. Web. April 15, 2014.

Emeljanow, Victor. "The Actor-Manager and the Iconoclast: Beerbohm Tree Produces Ibsen's *An Enemy of the People*." *On-Stage Studies* 20 (1997): 100-16. Print.

An Enemy of the People. Adap. Arthur Miller. Dir. Jack O'Brien. Perf. John Glover, George Grizzard, Valerie Mahaffey, William Anton. American Playhouse, 1990. Print.

Engel, Richard. "The Pornography of War." Blogging Baghdad: The Untold Story. MSNBC. April 21, 2006. Web. June 5, 2014.

Erin Brokovich. Dir. Steven Soderbergh. Perf. Julia Roberts and Albert Finney. Universal, 2000.

Estok, Simon C. "Shakespeare and Ecocriticism: An Analysis of 'Home' and 'Power' in *King Lear*." *AUMLA* 103 (May 2005): 15-41. Print.

Evernden, Neil. *The Social Creation of Nature*. Baltimore: Johns Hopkins University Press, 1992. Print.

Farmer, Jared. *Glen Canyon Dammed: Inventing Lake Powell and the Canyon Country*. Tucson: University of Arizona Press, 1999. Print.

Faulkner, William. *Light in August*. New York: Modern Library, 1932. Print.

Featherstone, Roger. "Earth First! & the IWW: An Interview with Roger Featherstone." *Industrial Worker* May 1998. Web. March 2, 2001.

"Fellow Workers, Meet Earth First!: An Open Letter to Wobblies Everywhere." *Industrial Workers of the World*. May 1998. Web. Sept. 13, 2013.

Fight Club. Dir. David Fincher. Perf. Brad Pitt, Edward Norton, Helena Bonham Carter. Fox, 2000. Film.

Flynn, Elizabeth Gurley. *Sabotage*. Cleveland, Ohio: IWW Publishing, 1916. Print.

Foreman, Dave. *Confessions of an Eco-Warrior*. New York: Crown, 1991. Print.

———, and Bill Haywood. *Ecodefense: A Field Guide to Monkeywrenching*. Second Edition. Tucson: Ned Ludd, 1987. Print.

Fresonke, Kris. *West of Emerson: The Design of Manifest Destiny*. Berkeley: University of California Press, 2003. Print.

Frick, George Frederick, and Raymond Phineas Stearns. *Mark Catesby: The Colonial Audubon*. Urbana: University of Illinois Press, 1961. Print.
Friedman, Mitch, and Paul Lindholdt, eds. *Cascadia Wild: Protecting An International Ecosystem*. Portland, OR: Frontier, 1993. Print.
Fumento, Michael. *Science under Siege: Balancing Technology and the Environment*. New York: Morrow, 1993. Print.
Gardner, Eric. "Producer Launches Legal War over New Eco-terrorist Film Starring Jesse Eisenberg." *The Hollywood Reporter* September 14, 2012. August 27, 2013. Web.
Garrard, Greg. *Ecocriticism*. 2nd ed. London: Routledge, 2012. Print.
Gates, Jr., Henry Louis. *Loose Canons: Notes on the Culture Wars*. New York: Oxford University Press, 1992. Print.
Gehrman, Elizabeth. *Rare Birds: The Extraordinary Tale of the Bermuda Petrel and the Man who Brought It Back from Extinction*. Boston: Beacon, 2012. Print.
Glendinning, Chellis. *When Technology Wounds: The Human Consequences of Progress*. New York: William Morrow, 1990. Print.
———. "Notes Toward a Neo-Luddite Manifesto." *Utne Reader* March/April 1990: 50-51. Print.
Goodman, Jordan, and Vivien Walsh. *The Story of Taxol: Nature and Politics in the Pursuit of an Anti-Cancer Drug*. New York: Cambridge University Press, 2001. Print.
Grumbling, Vernon Owen. "Chapter 7: Literature." *Greening the College Curriculum: A Guide to Environmental Teaching in the Liberal Arts*. Ed. Jonathan Collett and Stephen Karakashian. Washington, DC: Island, 1996. Print.
Hall, David D. *Worlds of Wonder, Days of Judgment: Popular Religious Belief in Early New England*. New York: Knopf, 1989. Print.
Hampton, Wayne. *Guerilla Minstrels: John Lennon, Joe Hill, Woody Guthrie, and Bob Dylan*. Knoxville: University of Tennessee Press, 1986. Print.
Harden, Blaine. *A River Lost: The Life and Death of the Columbia*. New York: Norton, 1996. Print.
Hardin, Garrett. "The Tragedy of the Commons." *Science* 162 (1968): 1243-1248. Print.
Harriot, Thomas. *A Brief and True Report of the New Found Land of Virginia*. Electronic Edition. Documenting the American South. University of North Carolina at Chapel Hill. 2004. Web. May 14, 2014.
Harrison, Robert Pogue. *Forests: The Shadow of Civilization*. Chicago: University of Chicago Press, 1992. Print.
Hawthorne, Nathaniel. "My Kinsman, Major Molineux." Electronic Text Center, University of Virginia Library, 1996. Web. May 7, 2014.
———. "Young Goodman Brown." *Nathaniel Hawthorne's Tales*. Norton Critical Edition. 2nd ed. Ed. James McIntosh. New York: Norton, 2013. Print.
Hegarty, Emily. "Dirty Pictures, Dirty Places: Environmentalism and Pornography." Center for Literary and Cultural Studies, Harvard University, Cambridge, MA. March 1996. Web (dead link).
Helvarg, David. *The War Against the Greens: The "Wise-Use" Movement, the New Right, and Anti-Environmental Violence*. San Francisco: Sierra Club, 1994. Print.
———. "The Anti-Enviro Connection." *The Nation* 260 (May 22, 1995): 722, 724. Print.
Hepworth, James, and Gregory McNamee, eds. *Resist Much, Obey Little: Some Notes on Edward Abbey*. Salt Lake City: Dream Garden, 1985. Print.
Herndon, Jerry A. "'Moderate Extremism': Edward Abbey and 'The Moon-Eyed Horse.'" *Western American Literature* 16.2 (August 1981): 97-103. Print.
Hill, Joe. "Ta-Ra-Ra-Boom-De-Ay!" http://www.daypoems.net/poems/2477.html. Web.
History of Lake Powell. Friends of Lake Powell, Inc. July 18, 2007. Web. June 2, 2014.
Howells, William Dean. "The Later Novel." *Cambridge History of American Literature*. Vol. 3. Later National Literature, Part II. Ed. William Peterfield Trent, John Erskine, Stuart P. Sherman, Carl Van Doren. New York: Macmillan, 1921. 66-95. Print.
Hughes, Robert. *The Wilderness and the West*. Dir. and prod. Kent James. Alexandria, VA: American Visions, 1996.

Hunt Saboteurs Association. 2014. Web. May 5, 2014.
Ibsen, Henrik. *An Enemy of the People*. 1882. Ed. Arthur Miller. New York: Penguin, 1951. Print.
If a Tree Falls: The Story of the Earth Liberation Front. Dir. Marshall Curry. Oscilloscope Laboratories, 2012. DVD.
"International Opportunities for Artists." *Leonardo* 7.1 (Winter 1974): 69. Print.
Irmscher, Christoph. *The Poetics of Natural History: From John Bartram to William James*. New Brunswick: Rutgers University Press, 1999. Print.
Irving, Washington. *A Tour on the Prairies*. 1835. Norman: University of Oklahoma Press, 1956. Print.
Jacobs, Lynn. *Waste of the West: Public Lands Ranching*. Tucson: P.O. Box 5784, 1991. Print.
Jensen, Derrick, and George Draffan. *Railroads and Clearcuts: Legacy of Congress's 1864 Northern Pacific Railroad Land Grant*. Spokane, WA: Inland Empire Public Lands Council, 1995. Print.
Johnston, Brian. "The Poetry of 'An Enemy of the People'." *Scandinavica: An International Journal of Scandinavian Studies* 18 (1979): 109-22. Print.
Jones, Steven E. *Against Technology: From the Luddites to Neo-Luddism*. New York: Routledge, 2006. Print.
Josselyn, John. *New-Englands Rarities Discovered*. London, 1672. Print.
———. *John Josselyn, Colonial Traveler: A Critical Edition of "Two Voyages to New-England."* 1674. Ed. Paul J. Lindholdt. Hanover, NH: University Press of New England, 1988. Print.
Kahn, Paul W. "American Exceptionalism, Popular Sovereignty, and the Rule of Law." *American Exceptionalism and Human Rights*. Ed. Michael Ignatieff. Princeton, NJ: Princeton University Press, 2005. 198-222. Print.
Kalm, Peter. *Peter Kalm's Travels in America: The English Version of 1770*. Ed. Adolph B. Benson. New York: Dover, 1987. Print.
Kastner, Joseph. *A Species of Eternity*. New York: Dutton, 1977. Print.
Keats, John. "Ode on a Grecian Urn." Poetry Foundation. 2014. Web. Sept. 11, 2014.
Kelly, Franklin. *Frederic Edwin Church and the National Landscape*. Washington, DC: Smithsonian, 1988. Print.
Kilian, Michael. "Turning Bureaucracy into Art Form: Interior's Mural Explains Mission with Symbolism." *Chicago Tribune* Aug. 21, 2003: 7. Print.
Kinsey, Joni Louise. *Thomas Moran and the Surveying of the American West*. Washington, DC: Smithsonian Institution, 1992. Print.
Kennedy, Robert F., Jr. "Our Environmental Destiny." Horizons Speakers Series. Eastern Washington University. Feb. 10, 1999.
Knighton, José. "Eco-porn and the Manipulation of Desire." *Wild Earth* 3.1 (Spring 1993): 76-78. Print.
Knutson, Harold C. *An Enemy of the People*: Ibsen's Reluctant Comedy." *Comparative Drama* 27.2 (Summer 1993): 159-75. Print.
Kohn, Stephen M. *Concepts and Procedures in Whistleblower Law*. Westport, CT: Quorum, 2001. Print.
Kraft, Michael E., and Diana Wuertz. "Environmental Advocacy in the Corridors of Government." In James G. Cantrill and Christine L. Oravec, eds. *The Symbolic Earth*, 95-122. Print.
Krawczynski, Keith T. *Daily Life in the Colonial City*. Portsmouth, NH: Greenwood, 2013. Print.
Krech III, Shepard. "Beyond *The Ecological Indian*." *Native Americans and the Environment: Perspectives on the Ecological Indian*. Ed. Michael E. Harkin and David Rich Lewis. Lincoln: University of Nebraska Press, 2007. 4-31. Print
Lake Powell . . . Jewel of the Colorado. Dir. W. L. Rusho. Prod. Bureau of Reclamation, in cooperation with the National Park Service, 1965.
Lakoff, George. *Don't Think of an Elephant*. White River Junction, VT: Chelsea Green, 2005. Print.

Lambert, Robert G. "An Enemy of the People: A Friend of the Teacher." *The English Journal* 54 (1965): 626-28. Print.
Lame Deer, John (Fire), and Richard Erdoes. *Lame Deer, Seeker of Visions*. New York: Washington Square, 1972. Print.
Lanham, Richard A. *Revising Prose*. 1979. 3rd ed. Boston: Allyn & Bacon, 1992. Print.
Lape, Noreen Groover. *West of the Border: The Multicultural Literature of the Western American Frontiers*. Athens: Ohio University Press, 2000. Print.
Lawrence, Claire. "'Getting the Desert into a Book': Nature Writing and the Problem of Representation in a Postmodern World." *Coyote in the Maze: Tracking Edward Abbey in a World of Words*. Ed. Peter Quigley. Salt Lake City: University of Utah Press, 1998. 150-167. Print.
Lemay, J. A. Leo. "The Frontiersman from Lout to Hero: Notes on the Significance of the Comparative Method and the Stage Theory in Early American Literature and Culture." *Proceedings of the American Antiquarian Society* 88 (Oct. 1979): 187-223. Print.
Leopold, Aldo. *Sand County Almanac*. 1949. New York: Ballantine, 1966. Print.
Levett, Christopher. *A Voyage into New England*. London, 1628. Print.
Lewis, Meriwether. "September 13, 1804." *The Journals of Lewis and Clark*. Project Gutenberg. Jan. 26, 2013. Web. Apr. 27, 2013.
Lewis, Thomas A. "Cloaked in a Wise Disguise." *Let the People Judge: Wise Use and the Private Property Rights Movement*. Ed. John Echeverria and Raymond Booth Eby. Washington, DC: Island, 1995. 13-20. Print.
Lindholdt, Paul J. "Counterscience and Conservation." *Skeptic* 5.1 (1997): 64-70. Print.
———. "Wise Use Joins the Militia." *Cascadia Times* 1.3 (June 1995): 1, 10-13. Print.
———. *In Earshot of Water: Notes from the Columbia Plateau*. Iowa City: University of Iowa Press, 2011. Print.
Loeffler, Jack. "Edward Abbey, Anarchism and the Environment." *Western American Literature* 28.1 (May 1993): 43-49. Print.
Lopez, Barry. "We Are Shaped by the Sound of Wind, the Slant of Sunlight." *High Country News* Sept. 14, 1998: 1, 4-5. Print.
"Lost Childhood: Russian Doukhobors or Sons of Freedom." *16x9 Global News*. Dec. 8, 2012. Web. May 2, 2014.
Lowitt, Richard. Review of *Colony and Empire: The Capitalist Transformation of the American West*, by William G. Robbins, in *Pacific Historical Review* 65 (Feb. 1996): 131-2. Print.
Lyon, Thomas. *This Incomperable Lande: A Book of American Nature Writing*. Boston: Houghton Mifflin, 1989. Print.
"Management of Artwork: Steps Taken to Preserve and Protect Bureau of Reclamation's Collection." *Federal Industry Watchdog* July 1, 1992. Retrieved via Westlaw, May 16, 2007. Web. June 1, 2014.
Mander, Gerry. "Ecopornography: One Year and Nearly a Billion Dollars Later, Advertising Owns Ecology." *Communication and Arts* 14.2 (1972): 45-56. Print.
Manes, Christopher. *Green Rage: Radical Environmentalism and the Unmaking of Civilization*. Boston: Little Brown, 1990. Print.
Marie, Lorraine. "'Black' helicopters bear 'Big Brother.'" Colville *Statesman-Examiner* Feb. 22, 1995: A1. Print.
Marston, Ed. "Floyd Dominy: An Encounter with the West's Undaunted Dam-builder." *High Country News* Aug. 28, 2000. Retrieved via Westlaw, June 5, 2007. Web. April 2, 2008.
Marx, Karl. *Capital: A Critique of Political Economy*. Vol. 1. Trans. Ben Fowkes. New York: Vintage, 1977. Print.
Marx, Leo. *The Machine in the Garden: Technology and the Pastoral Ideal in America*. 1964. New York: Oxford University Press, 2000. Print.
McCloskey, David. "Ecology and Community: The Bioregional Vision." Self-published essay, 1991. Web (dead link).
McCourt, Frank. *Angela's Ashes: A Memoir*. New York: Scribner's, 1996. Print.

McFarlane, Robert. "The Prophet in the Wilderness." *Wall Street Journal* May 3, 2013. Web. May 27, 2013.

McPhee, John. *Encounters with the Archdruid*. New York: Farrar, Straus and Giroux, 1971. Print.

Mentz, Steve. "After Sustainability." *PMLA* 127.3 (May 2012): 586-92. Web. Apr. 23, 2014. Print.

Miller, Arthur. "Ibsen's Warning." *Index on Censorship* 18 (July-Aug. 1989): 74-76. Print.

———. *Arthur Miller's Adaptation of An Enemy of the People*. 1950. New York: Penguin, 1977. Print.

Moffett, Marian, and Lawrence Wodehouse. *Built for the People of the United States: Fifty Years of TVA Architecture*. Knoxville: Art and Architecture Gallery, University of Tennessee Press, 1983. Print.

Mooallem, Jon. *Wild Ones: A Sometimes Dismaying, Weirdly Reassuring Story about Looking at People Looking at Animals in America*. New York: Penguin, 2013. Print.

Morison, Samuel Eliot. *The European Discovery of America: The Northern Voyages*. New York: Oxford University Press, 1971. Print.

Moyers & Company. "Richard Slotkin on Guns and Violence." Dec. 13, 2013. Web. Jun. 8, 2014.

Mumford, Lewis. *The Myth of the Machine*. Vol. 1: *Technics and Human Development*. New York: Harcourt, Brace & World, 1967. Vol. 2: *The Pentagon of Power*. New York: Harcourt Brace Jovanovich, 1970. Print.

Murphy, Patrick D. *Farther Afield in the Study of Nature-Oriented Literature*. Charlottesville, VA: University of Virginia Press, 2000. Print.

NPR: National Public Radio. "How Could a Drought Spark a Civil War?" Sept. 8, 2013. Web.

Nash, Gerald D. *The Federal Landscape: An Economic History of the Twentieth-Century West*. Tucson: University of Arizona Press, 1999. Print.

Nash, Roderick. *Wilderness and the American Mind*. New Haven: Yale University Press, 1967. Print.

———. *The Rights of Nature: A History of Environmental Ethics*. Madison: University of Wisconsin Press, 1989. Print.

National Park Service. "Yellowstone National Park: Denning and Hibernation Behavior." March 31, 2014. Web. April 15, 2014.

Night Moves. Dir. Kelley Reichardt. Perf. Dakota Fanning, Jesse Eisenberg, Alia Shawkat. Cinedigm, 2013. Film.

Novak, Barbara. *Nature and Culture: American Landscape and Painting, 1825-1875*. New York: Oxford University Press, 1980. Print.

Noble, David F. *Progress without People: In Defense of Luddism*. Chicago: Charles H. Kerr, 1993. Print.

Novak, Barbara. *Nature and Culture: American Landscape and Painting, 1825-1875*. New York: Oxford University Press, 1980. Print.

Okanogan National Forest. "Livestock Use By Allotment." 1989. Print.

Oliveria, D. F. "Inside Hate's Gate." [Spokane] *Spokesman-Review* May 27, 2001: B8. Print.

Olympic National Park, Washington. "Frequently Asked Questions." National Park Service. May 21, 2014. Web. May 22, 2014.

Opel, Andy. "Corporate Culture Keeps Nature Regular: The 'Super Citizen,' the Media and the 'Metamucil and Old Faithful' Ad." *Capitalism, Nature, Socialism* 17.3 (Sept. 2006): 100-13. Print.

Oravec, Christine L. "To Stand Outside Oneself: The Sublime in the Discourse of Natural Scenery." In James G. Cantrill and Christine L. Oravec, eds. *The Symbolic Earth*, 58-75. Print.

Orr, David W. "Reinventing Higher Education." In *Greening the College Curriculum: A Guide to Environmental Teaching in the Liberal Arts*. Eds. Jonathan Collett and Stephen Karakashian. Washington, DC: Island, 1996. 8-23. Print.

———. *Earth in Mind: On Education, Environment, and the Human Prospect*. Washington, DC: Island, 1994. Print.

Pally, Marc. "The Enterprise of Process: Notes on Planning for Public Art." National Endowment for the Arts. Feb. 11, 2014. Web. June 2, 2014.

Parini, Jay. "Regionalism and Literary Studies." *The Chronicle of Higher Education* Jan. 17, 1997: 60. Print.

Partridge, Ernest. "Disequilibrium Ecology – Much Ado about Nothing." Paper read March 23, 2005, Pacific Division Meeting, International Society for Environmental Ethics. Web. April 26, 2014.

Payne, Daniel G. "Monkey Wrenching, Environmental Extremism, and the Problematical Edward Abbey." *Southwestern American Literature* 21.1 (Fall 1995): 195-208. Print.

Paystrup, Patricia. "Plastics as a 'Natural Resource': Perspective by Incongruity for an Industry in Crisis." In James G. Cantrill and Christine L. Oravec, eds. *The Symbolic Earth*, 176-97. Print.

Perry, Lewis. *Civil Disobedience: An American Tradition*. New Haven, CT: Yale University Press, 2013. Print.

Pitzer, Paul C. *Grand Coulee: Harnessing a Dream*. Pullman: Washington State University Press, 1994. Print.

Plotkin, Cary H. "Towards a Poetics of Transcendence after Charles Darwin." Gerard Manley Hopkins Archive. June 11, 2011.

Pombo, Richard, and Joseph Farah. *This Land is Our Land: How to End the War on Private Property*. New York: St. Martin's, 1996. Print.

Ponting, Clive. *A Green History of the World: The Environment and the Collapse of Great Civilizations*. New York: Penguin, 1991. Print.

Prown, Jules David, ed. *Discovered Lands, Invented Pasts: Transforming Visions of the American West*. New Haven: Yale University Press, 1992. Print.

Pryne, Eric. "Park plan called twisted global plot to destroy nation." Spokane *Spokesman-Review* October 23: A1. Print.

Pring, George W., and Penelope Canan. *SLAPPs: Getting Sued for Speaking Out*. Philadelphia: Temple University Press, 1996. Print.

Proulx, Annie. *Close Range: Wyoming Stories*. New York: Scribner, 1999. Print.

———. *That Old Ace in the Hole*. New York: Scribner, 2002. Print.

Pynchon, Jr., Thomas Ruggles. "Is it O.K. to be a Luddite?" *The New York Times*. October 28, 1984. September 18, 2013. Web.

Radaker, Kevin. "Henry Thoreau and Frederic Church: Confronting the Monumental Sublimity of the Maine Wilderness." *Yearbook of Interdisciplinary Studies in the Fine Arts* 1 (1989): 267-88. Print.

Ramos, Tarso. "Wise Use in the West: The Case of the Northwest Timber Industry." Ed. John Echeverria and Raymond Booth Eby. *Let the People Judge: Wise Use and the Private Property Rights Movement*. Washington, DC: Island, 1995. 82-118. Print.

Rand, Ayn. *Introduction to Objectivist Epistemology*. 2nd ed. New York: Penguin, 1990. Print.

Ray, Dixie Lee. *Environmental Overkill: Whatever Happened to Common Sense?* Washington, DC: Regnery Gateway, 1993. Print.

Ray, Dixie Lee, and Lou Guzzo. *Trashing the Planet: How Science Can Help Us Deal with Acid Rain, Depletion of the Ozone, and Nuclear Waste (Among Other Things)*. Washington, DC: Regnery Gateway, 1990. Print.

Regis, Pamela. *Describing Early America: Bartram, Jefferson, Crevecoeur, and the Rhetoric of Natural History*. DeKalb: Northern Illinois University Press, 1992. Print.

Reisner, Marc. *Cadillac Desert: The American West and Its Disappearing Water*. New York: Viking Penguin, 1986. Print.

"Researchers track New Hampshire moose in hopes of pinpointing cause of population decline." PBS NewsHour. April 7, 2014. Web. April 15, 2014.

Richards, Kent D. "The Young Napoleons: Isaac Stevens, George McClellan and the Northern Railroad Survey." *Columbia* 3.4 (1989/90): 21-28. Print.

Ring, Ray. "Taking Liberties." *High Country News*. July 24, 2006: 8-14. Print.
Robbins, William G. *Colony and Empire: The Capitalist Transformation of the American West*. Lawrence: University of Kansas Press, 1994. Print.
Rockwell, Norman. *Glen Canyon Dam*. Bureau of Reclamation Fine Art Collection. Denver, CO. Sept. 13, 2013. Web. June 7, 2014.
Rogers, Pattiann. *Firekeeper: New and Selected Poems*. Minneapolis: Milkweed, 1994. Print.
Ronald, Ann. *The New West of Edward Abbey*. Albuquerque: University of New Mexico Press, 1982. Print.
Roorda, Randall. (1997). "Sites and Senses of Writing in Nature." *College English* 59 (1997): 385-407. Print.
Rosaldo, Renato. *Culture and Truth: The Remaking of Social Analysis*. 1989. Boston: Beacon, 1993. Print.
Rosowski, Susan J. *Birthing a Nation: Gender, Creativity, and the West in American Literature*. Lincoln: University of Nebraska Press, 1999. Print.
Rowell, Andrew. *Green Backlash: Global Subversion of the Environmental Movement*. London: Routledge, 1996. Print.
Ruby, Robert H., and John A. Brown. *A Guide to the Tribes of the Pacific Northwest*. Norman: University of Oklahoma Press, 1986. Print.
Rutledge, Chris R., and Terry McLendon. N.d. "An Assessment of Exotic Plant Species of Rocky Mountain National Park." Global Change Master Directory. NASA. April 2014. Web. April 28, 2014.
Ryberg, Erik. Personal interview, conducted May 30, 1993, in Dixie, Idaho.
———. Telephone interview. 26 January 1999.
"Sabots - toss your shoes into the machine works." Arete Designs. Sept. 18, 2013. Web (dead link).
Safina, Carl. "To Save the Earth, Scientists Should Join Policy Debates." *The Chronicle of Higher Education* Nov. 6, 1998: A80. Print.
"Sahara Club's Agenda to Attack Environmentalists." *Nature's Advocate: Newsletter of the Upper Columbia River Sierra Club* (Feb. 1995): 13. Print.
Sale, Kirkpatrick. *Dwellers in the Land: The Bioregional Vision*. San Francisco: Sierra Club, 1985. Print.
———. *Rebels Against the Future: The Luddites and Their War on the Industrial Revolution – Lessons for the Computer Age*. Reading, MA: Addison-Wesley, 1995. Print.
Scarce, Rik. *Eco-Warriors: Understanding the Radical Environmental Movement*. Chicago: Noble, 1990. Print.
Schultz, Stanley K. "American History 102: Civil War to the Present." January 20, 2000. Web. June 6, 2014.
Scott, J. Michael, Dale D. Goble, Aaron M. Haines, John A Wiens, and Maile C. Neele. "Conservation-Reliant Species and the Future of Conservation." *Conservation Letters* 3 (2010): 91-97. Web. June 3, 2014.
Seeger, Pete, and Lee Hays. "If I Had a Hammer." Metro Lyrics. Web. May 7, 2014.
Senate Bill 6566 (dead). Washington State Legislature. 2010 Regular Session. Web. May 12, 2014.
"Settling fish vs. dams: Is there a better time?" *Lewiston Tribune* Nov. 11, 2012. Web. April 28, 2014.
Shapiro, Ellen. "Greenwashing!" *Advertising Annual* (2004): 196-200. Print.
Sheehan, Bernard. Rev. of *Describing Early America: Bartram, Jefferson, Crevecoeur, and the Rhetoric of Natural History*, by Pamela Regis. *Journal of American History* 81.1 (Jun. 1994): 254-55. Print.
Silkwood. Dir. Mike Nichols. Perf. Meryl Streep, Cher, and Kurt Russell. New Line Home Video, 1983.
Sinclair, Upton. *The Jungle*. 1906. New York: Bantam, 1981. Print.
Slotkin, Richard. *Regeneration Through Violence: The Mythology of the American Frontier, 1600-1860*. Middletown, CT: Wesleyan University Press, 1973. Print.

———. *The Fatal Environment: The Myth of the Frontier in the Age of Industrialization.* New York: Atheneum, 1985. Print.
———. *Gunfighter Nation: The Myth of the Frontier in Twentieth-Century America.* New York: Atheneum, 1992. Print.
Slovic, Scott. *Seeking Awareness in American Nature Writing.* Salt Lake City: University of Utah Press, 1992. Print.
Smith, Gibbs M. *Labor Martyr Joe Hill.* New York: Grossett & Dunlap, 1969. Print.
Smith, John. *The Complete Works of Captain John Smith (1580-1631).* Ed. Philip L. Barbour. 3 Vols. Chapel Hill: University of North Carolina Press, 1986. Print.
Snyder, Gary. *The Practice of the Wild.* San Francisco: North Point, 1990. Print.
"Sons of Liberty." *Boston Tea Party: A Revolutionary Experience.* Massachusetts Convention Center Authority. 2014. Web. May 2, 2014.
Soulé, Michael, and Reed Noss. "Rewilding and Biodiversity: Complementary Goals for Continental Conservation." *Wild Earth* (Fall 1998): 1-11. Print.
Splawn, A. J. *Ka-mi-akin: Last Hero of the Yakimas.* 1917. Portland, OR: Binford & Mort, 1944. Print.
Sprouse, Martin, ed. *Sabotage in the American Workplace: Anecdotes of Dissatisfaction, Mischief and Revenge.* San Francisco: Pressure Drop, 1992. Print.
Stallwood, Kim. "A Personal Overview of Direct Action in the United Kingdom and the United States." In Best and Nocella, 81-90. Print.
Stearns, Raymond Phineas. *Science in the British Colonies of America.* Urbana: University of Illinois Press, 1970. Print.
Stegner, Wallace. *Joe Hill: A Biographical Novel.* New York: Penguin, 1950. Print.
———. "Striking the Rock." *The American West as Living Space.* Ann Arbor: University of Michigan Press, 1987. 29-60. Print.
———. "It All Began with Conservation." *Smithsonian* Apr. 1990. Retrieved via Westlaw, May 16, 2007. Web. May 12, 2014.
Steinbeck, John. *Grapes of Wrath and Other Writings, 1936-1941.* New York: Library of America, 1996. Print.
Stern, Kenneth S. *A Force upon the Plain: The American Militia Movement and the Politics of Hate.* New York: Simon & Schuster, 1996. Print.
Stevens, Wallace. "Anecdote of the Jar." *Harmonium.* New York: Alfred A. Knopf, 1923. Rpt. in *The Collected Poems of Wallace Stevens.* New York: Alfred A. Knopf, 1961. 44. Print.
Stewart, Mart A. Rev. of *Describing Early America: Bartram, Jefferson, Crevecoeur, and the Rhetoric of Natural History,* by Pamela Regis. *Forest and Conservation History* 37.3 (Jul. 1993): 142. Print.
Stiles, Jim. "Cantankerous and Contradictory: Remembering Ed Abbey." *High Country News.* Mar. 29, 1999. Web. May 12, 2014.
Stinger, Leslie, and Bobbie Ferguson. "Portraits of Reclamation." *Cultural Resources Management* 22.4 (1999): 48-50. Web. June 4, 2014.
Storey, Brit Allan. Senior Historian, Bureau of Reclamation. Telephone interview, June 3, 2007.
Strong, George Templeton. *The Diary of George Templeton Strong.* Vol. 2: 1850 - 1859. Ed. Allan Nevins and Milton Halsey Thomas. New York: Macmillan, 1952. Print.
Sugarman, Josh. "Smith & Wesson Snubs Police with 'Commemorative' Revolver." *Huffington Post.* July 22, 2008. Web. June 6, 2014.
Tall, Deborah. *From Where We Stand: Recovering a Sense of Place.* Baltimore: Johns Hopkins University Press, 1993. Print.
Thomashow, Mitchell. *Ecological Identity: Becoming a Reflective Environmentalist.* Cambridge: Massachusetts Institute of Technology, 1995. Print.
Thoreau, Henry David. *A Week on the Concord and Merrimack Rivers.* 1849. Boston: Houghton Mifflin, 1906. Print.
Timeline of Woody Guthrie (1912-1967). American Memory. The Library of Congress. July 10, 2007. Web. April 14, 2014.

Taxpayers for Common Sense. "Dam Removal is the Only Solution." Sept. 6, 2002. Web. July 20, 2007.
Todd, Anne Marie. "The Aesthetic Turn in Green Marketing: Environmental Consumer Ethics of Natural Personal Care Products." *Environmental Ethics* 9.2 (2004): 86-102. Print.
Trombold, John. "Popular Songs as Revolutionary Culture in John Dos Passos' *U.S.A.* and Other Early Works." *Journal of Modern Literature* 19.2 (Fall 1995): 289-316. Print.
Tuan, Yi-Fu. *Topophilia: A Study of Environmental Perceptions, Attitudes, and Values*. 1974. New York: Columbia University Press, 1990. Print.
Twining, Edward S. "Edward Abbey, American: Another Radical Conservative." *Denver Quarterly* 12.4 (1978): 3-15. Print.
U.S. Senate Committee on Environment and Public Works. Hearing on Conservation and Reinvestment Act Bills. Washington, D.C. May 24, 2000. Web. June 15, 2014.
"U.S. University Spying Scandal Prompts Resignations." *The Guardian*. Jan. 20, 2006. Web. Jan. 23, 2006.
Ulrich, Laurel Thatcher. *A Midwife's Tale: The Life of Martha Ballard, Based on Her Diary, 1785-1812*. New York: Random House, 1990. Print.
Vale and Valleys Hunt Saboteurs. *Sabotage*. 1992. Web. May 10, 2014.
"Valuable Art Missing from Interior Dept." *San Francisco Chronicle*. Aug. 18, 1990. Westlaw. May 16, 2007.
Van Laan, Thomas F. "Generic Complexity in Ibsen's *An Enemy of the People*." *Comparative Drama* 20.2 (Summer 1986): 95-114. Print.
Veasaw, Kendall. "How to Organize Eco Terrorists: The Ironic Effect of the 'Green Scare.'" Capstone Paper at Siena College. Nov. 12, 2013. Web. May 6, 2014.
Verrill, Addison Emery. "The Story of the Cahow: The Mysterious Extinct Bird of the Bermudas." *Popular Science Monthly* 60 (Nov. 1901). Jan. 25, 2014. Web. Apr. 23, 2014.
Walter, Jess. *Ruby Ridge: The Truth and Tragedy of the Randy Weaver Family*. New York: Regan, 2000. Print.
Wardenaar, Leslie A. "Humor in the Colonial Promotion Tract: Topics and Techniques." *Early American Literature* 9 (1975): 286-300. Print.
Welchman, Jennifer. "Is Ecosabotage Civil Disobedience?" *Philosophy and Geography* 4.1 (2001): 97-107. Print.
Welling, Bart. "Ecoporn: On the Limits of Visualizing the Nonhuman." *Ecosee: Image, Rhetoric, and Nature*. Ed. Sidney I. Dobrin and Sean Morey. Albany: State University of New York Press, 2009. 53-78. Print.
Weltzien, Alan. *Exceptional Peaks*, forthcoming from University of Nebraska Press, 2015.
Wharton, Donald P. "John Josselyn (c. 1608-1675)." *American Writers Before 1800: A Biographical and Critical Dictionary*. Ed. James A. Levernier and Douglas R. Wilmes. 3 Vols. Westport, CT: Greenwood, 1983. Print.
Wilderness Act: The Wilderness Legislation of 1964. Jan. 8, 2004. Web. May 14, 2014.
Wilkinson, Charles F. *The Eagle Bird: Mapping a New West*. New York: Pantheon, 1992. Print.
Wilkinson, Todd. *Science Under Siege: The Politicians' War on Nature and Truth*. Boulder, CO: Johnson, 1998. Print.
Williams, John H. Introduction. *The Canoe and the Saddle; or Klalam and Klickitat, to Which Are Now First Added His Western Letters and Journals*. Ed. John H. Williams. Tacoma, WA: John H. Williams, 1913. Print.
Williams, Terry Tempest. *Refuge: An Unnatural History of Family and Place*. New York: Vintage, 1992. Print.
Wilson, Richard Guy. "Machine-Age Iconography in the American West: The Design of Hoover Dam." *Pacific Historical Review* 54.4 (1985): 463-93. Print.
Wilson, David Scofield. *In the Presence of Nature*. Amherst: University of Massachusetts Press, 1978. Print.

Wilson, E. O. *The Diversity of Life*. Cambridge, MA: Harvard University Press, 1984. Print.

Wilton, Andrew, and Tim Barringer. *American Sublime: Landscape Painting in the United States, 1820–1880*. Princeton: Princeton University Press, 2002. Print.

Winthrop, Theodore. *A Companion to the Heart of the Andes*. New York: Appleton, 1859. Print.

———. *The Canoe and the Saddle, Adventures among the Northwestern Rivers and Forests, and Isthmiana*. Boston: Ticknor and Fields, 1862. Print.

———. *The Canoe and the Saddle: A Critical Edition*. Ed. Paul J. Lindholdt. Lincoln: Bison–University of Nebraska Press, 2006. Print.

Wood, William. *New-England's Prospect*. London, 1634. Print.

Woolf, Eugene T. *Theodore Winthrop: Portrait of an American Author*. Washington D.C.: University Press of America, 1981. Print.

Worster, Donald. Rev. of *Cadillac Desert: The American West and Its Disappearing Water*. *Western Historical Quarterly* 18.3 (1987): 346-47. Print.

Wordsworth, William. "Preface to *Lyrical Ballads* (1800)." *The Prose Works of William Wordsworth*. Vol. 1. Ed. W. J. B. Owen and Jane Worthington Smyser. Oxford: Clarendon, 1974. 116-158. Print.

Wrigley, Robert. "Babel." *Anatomy of Melancholy and Other Poems*. New York: Penguin, 2013. 54-55. Print.

Zakin, Susan. *Coyotes and Town Dogs: Earth First! and the Environmental Movement*. New York: Viking Penguin, 1993. Print.

Zisk, Betty H. Rev. of *Images of American* Radicalism, by Paul Buhle and Edmund B. Sullivan. *Peacework*. Feb. 1999. N.d. Web. May 6, 2014.

Zoeckler, John "A.P. Environmental Science: Research on Ecopornography." Jan. 6, 2004. Web (dead link).

Index

About the Author

Paul Lindholdt, a native of Seattle, worked throughout college as a homebuilder, industrial waste technician, Teamster, and longshoreman. His writing has won awards from the Academy of American Poets and the Society of Professional Journalists. In 2012 he received the Washington State Book Award in Biography/Memoir for *In Earshot of Water*. He lives in Spokane with his wife and their two boys, and is professor of English at Eastern Washington University.

www.ingramcontent.com/pod-product-compliance
Ingram Content Group UK Ltd.
Pitfield, Milton Keynes, MK11 3LW, UK
UKHW041125071026
15587UKWH00009B/207

9 780739 194980